Stewardship Planet

This volume examines historical views of stewardship that have sometimes allowed humans to ravage the earth as well as contemporary and futuristic visions of stewardship that will be necessary to achieve pragmatic progress to save life on Earth as we know it.

The idea of stewardship – human responsibility to tend the earth – has been central to human cultures throughout history, as evident in the Judeo-Christian Genesis story of the Garden of Eden and in a diverse range of parallel tales from other traditions around the world. Despite such foundational hortatory stories about preserving the earth on which we live, humanity in the Anthropocene is nevertheless currently destroying the planet with breathtaking speed.

Much research on stewardship today – in the disciplines of geography, urban studies, oceans research, and green business practice – offers insights that should help address the ecological challenges facing the planet. Simultaneous scholarship in the humanities and other fields reminds us that the damage done to the planet has often been carried out in the name of tending the land. In order to make progress in environmental stewardship, scholars must speak to each other across the disciplinary boundaries, as they do in this volume.

Rachel Carnell is Professor of English at Cleveland State University. Having published extensively on eighteenth-century literature and politics, she began working on environmental stewardship after unearthing archival references to eighteenth-century lawsuits that described landed estates in terms of expected monetary output.

Chris Mounsey is Professor of Eighteenth-century Cultural Studies at the University of Winchester. He has published widely on a range of issues including sexuality, disability, and bioethics. He is series editor of the Routledge Advances in the History of Bioethics.

Routledge Advances in the History of Bioethics

Routledge Advances in the History of Bioethics aims to act as a nexus for debates typically in collections of diverse but explicitly interrelated essays about the histories and literatures of bioethical debates from a wide spectrum of disciplines, methodologies, periods and geographical contexts. This series champions conversations from within interdisciplinary collision spaces, considering the effects of physical and metaphysical environments upon factual and fictional spaces.

Series Editors: Chris Mounsey, Stan Booth, and Madeleine Mant

Bodies of Information
Reading the Variable Body from Roman Britain to Hip Hop
Edited by Chris Mounsey and Stan Booth

The History and Bioethics of Medical Education
"You've Got to Be Carefully Taught"
Edited by Madeleine Mant and Chris Mounsey

Reconsidering Extinction in Terms of the History of Global Bioethics
Edited by Stan Booth and Chris Mounsey

Stewardship and the Future of the Planet
Promise and Paradox
Edited by Rachel Carnell and Chris Mounsey

For more information about this series, please visit: https://www.routledge.com/Routledge-Advances-in-the-History-of-Bioethics/book-series/RAITHOB

Stewardship and the Future of the Planet
Promise and Paradox

Edited by
Rachel Carnell and Chris Mounsey

NEW YORK AND LONDON

First published 2023
by Routledge
605 Third Avenue, New York, NY 10158

and by Routledge
4 Park Square, Milton Park, Abingdon, Oxon, OX14 4RN

Routledge is an imprint of the Taylor & Francis Group, an informa business

© 2023 selection and editorial matter, Rachel Carnell and Chris Mounsey; individual chapters, the contributors

Excerpts in Chapter 3 from "Energy Is Eternal Delight", "Four Changes", and "Magpie's Song" by Gary Snyder, from TURTLE ISLAND, copyright © 1974 by Gary Snyder. Reprinted by permission of New Directions Publishing Corp.

The right of Rachel Carnell and Chris Mounsey to be identified as the authors of the editorial material, and of the authors for their individual chapters, has been asserted in accordance with sections 77 and 78 of the Copyright, Designs and Patents Act 1988.

All rights reserved. No part of this book may be reprinted or reproduced or utilised in any form or by any electronic, mechanical, or other means, now known or hereafter invented, including photocopying and recording, or in any information storage or retrieval system, without permission in writing from the publishers.

Trademark notice: Product or corporate names may be trademarks or registered trademarks, and are used only for identification and explanation without intent to infringe.

Library of Congress Cataloging-in-Publication Data
A catalog record for this title has been requested

ISBN: 978-1-032-11245-9 (hbk)
ISBN: 978-1-032-11251-0 (pbk)
ISBN: 978-1-003-21906-4 (ebk)

DOI: 10.4324/9781003219064

Typeset in Sabon
by SPi Technologies India Pvt Ltd (Straive)

Contents

List of Illustrations	vii
Acknowledgments	viii
List of Contributors	ix

Introduction: The Contradictory Inflections of Stewardship   1
RACHEL CARNELL

PART I
Human Self-Perception and Misperception   11

1 Stewardship and Sense of Place: Assumptions and Ideals   13
TYRA A. OLSTAD

2 "I Was Under No Necessity of Seeking My Bread": *Robinson Crusoe* and the Stewardship of Resources in Eighteenth-Century England   29
CHRIS MOUNSEY

3 Stewardship in American Literature: Promise and Paradox in the New World   50
JOSH A. WEINSTEIN

PART II
Dystopian Visions of Past, Present, and Future   73

4 Monstrous Stewardship and the Plantation in Charles Chesnutt's "The Goophered Grapevine"   75
MATTHEW WYNN SIVILS

5 Human Stewardship and "Reproductive Futurism" in Dystopian Fiction   86
PRAMOD K. NAYAR

6 Climate Change and Apocalyptic Literature: Post-Human Stewardship in Paolo Bacigalupi's *Drowned Cities* Trilogy   104
JEFF KAREM

PART III
Approaches to Contemporary Challenges   133

7 Political Aspects of Stewardship for Wildlife in the U.S.   135
BRUCE ROCHELEAU

8 The Future of the Seascape and the Humanity of Islanders: Focusing on the Korean Archipelago   156
SUN-KEE HONG

9 Stewardship of Rangelands in the 21st Century: Managing Complexity from the Margins   175
NATHAN F. SAYRE

PART IV
Envisioning the Future   193

10 Product Stewardship: Ethics and Effectiveness in a Circular Economy   195
HELEN LEWIS AND NICK FLORIN

11 An Evolutionary Systems Theoretic Perspective on Global Stewardship   212
WILLIAM M. BOWEN

12 Stewardship in the Anthropocene: Meanings, Tensions, Futures   234
MARIA TENGÖ, JOHAN ENQVIST, SIMON WEST, UNO SVEDIN, VANESSA A. MASTERSON, AND L. JAMILA HAIDER

Index   252

Illustrations

Figures

10.1 Key differences between EPR and product stewardship   196
12.1 Stewardship as a boundary object based on three
     connecting dimensions – care, knowledge and agency   242

Tables

8.1 Research topic and academic fields, focusing on keywords
    available for collaboration   165
12.1 Four broad meanings of stewardship in the academic
     literature from 1990 to 2016   237

Acknowledgments

This volume would not have been possible without the support of a 2019 Faculty Scholarship Initiative Grant from Cleveland State University's Office of Research. The research funded by that grant allowed Rachel Carnell not only to finish a book on impeachment in the reign of Queen Anne, but also to stumble by chance on a document in the Manuscripts Collection at the British Library that sparked a conversation with Chris Mounsey about eighteenth-century notions of land stewardship. That conversation continued long-distance and eventually extended to ideas about stewardship across a broad range of disciplines. We are grateful to the scholars in these other fields around the globe for contributing their research to this volume (and for having patience with Covid-induced publishing delays). We likewise extend our gratitude to the University of Winchester for continuing support for Chris Mounsey's research, and also to the British Government's Access to Work Scheme. We thank Stan Booth for his ongoing technical advice and guidance. Rachel Carnell also extends her thanks to Jeff Karem, Jerzy Sawicki, and Lori O'Laughlin at Cleveland State for their continued support of her research endeavors – and to Greg and Alison Lupton for their enthusiastic interest in this project.

Contributors

William M. Bowen holds the position of Professor of Public Administration and Urban Studies at Cleveland State University. His interest in systems theory began while he was a Zoology undergraduate student studying ecology in the 1970s at the University of North Carolina at Chapel Hill. This interest was developed further while he was a Supply Officer in the U.S. Navy during the 1980s, and still further in his Master of Public Administration degree program at the College of Charleston and PhD in Regional Analysis and Planning at Indiana University in Bloomington. His research focus is on decision-making in public and environmental affairs, economic development, energy policy, and higher education. Throughout his career he has viewed stewardship of culture, knowledge, and ecological systems as the motivating factor behind virtually all his research and scholarship.

Rachel Carnell is Professor of English at Cleveland State University. She is the author of *Backlash: Libel, Impeachment, and Populism in the Reign of Queen Anne* (2020); *A Political Biography of Delarivier Manley* (2008); and *Partisan Politics, Narrative Realism, and the Rise of the British Novel* (2006). She is the co-editor of *The Secret History in Literature, 1660–1820* (2017) and of *The Selected Works of Delarivier Manley* (2005). She began working on environmental stewardship after unearthing archival references to eighteenth-century lawsuits that described landed estates in terms of expected monetary output.

Johan Enqvist completed his PhD at the Stockholm Resilience Centre (SRC, Stockholm University) in 2017, on stewardship and the role of local communities in managing urban environments in South Asia and North America. His subsequent research with the SRC and the African Climate & Development Initiative (University of Cape Town) has explored how Cape Town and its residents cope and adapt to an existential environmental threat in the form of the "Day Zero drought" and its aftermath. He is increasingly interested in how sustainability and justice intersect in the daily lives of urban dwellers, as well as contestations of what kinds of nature "belong" in cities and other human-dominated landscapes.

Nick Florin is a Research Director at the Institute for Sustainable Futures (ISF), University of Technology, Sydney. He leads the Resource Futures research group and directs and undertakes collaborative research projects with industry and government. His research involves policy options analysis, waste management technology and infrastructure assessment, material flow analysis, and stakeholder engagement to improve policy relevance and support sustainable supply chains. Nick's recent work has focused on reuse and recycling of photo voltaic panels, batteries and sustainable packaging.

L. Jamila Haider, PhD, is a researcher at the Stockholm Resilience Centre, and leader of the Resilience & Sustainable Development research theme. She studies the relationships between cultural and biological diversity through the lens of food. Her current research focuses on how to reconceptualize development pathways from a social-ecological systems perspective. As part of a new generation of sustainability scientists, and as a member of numerous early-career professional networks, Jamila teaches and writes about research journeys and how to enact transformational leadership with care. Jamila has worked as a development practitioner in Central Asia and Afghanistan and is author of the award-winning book *With Our Own Hands: A Celebration of Food and Life in the Pamir Mountains of Afghanistan and Tajikistan*.

Sun-Kee Hong received a PhD from Hiroshima University in Japan for research on landscape ecosystems in rural and suburban Asia. He has been a professor at Mokpo National University since 2005, doing research on the islands of Korea and Southeast Asia. Currently, he is a professor at the Institution for Marine and Island Cultures and an adjunct professor at the Department of Liberal Arts in Mokpo National University. Sun-Kee is editor or co-editor of many books, including *Ecological Issues in a Changing World* (2004), *Landscape Ecological Applications in Man-Influenced Areas: Linking Man and Nature Systems* (2007), *Landscape Ecology in Asian Cultures* (2011), *Biocultural Landscapes* (2014), *Landscape Ecology for Sustainable Society* (2018), and *Conserving Biocultural Landscape in Malaysia and Indonesia for Sustainable Development* (in press). He is editor-in-chief and founder of the *Journal of Marine and Island Cultures*.

Jeff Karem is Professor of English and Africana Studies at Cleveland State University. His research and teaching focus upon twentieth- and twenty-first-century American literatures, with an emphasis on migration and cultural interconnection throughout the Americas. Karem has published articles on regional and ethnic literatures throughout the Americas and is the author of two books: *The Romance of Authenticity:*

*The Cultural Politics of Regional and Ethnic Literature*s (2004) and *The Purloined Islands: Caribbean–U.S. Cross-Currents in Literature and Culture* (2011).

Helen Lewis has worked in the environmental management field for over 30 years and is an Adjunct Professor at the Institute for Sustainable Futures (ISF), University of Technology, Sydney. She currently works as an environmental consultant specializing in product stewardship and sustainable packaging. Helen also works for a variety of clients in government and the private sector to promote design for sustainability and increased recovery of products and packaging at end of life. Helen has been actively involved in the development of two product stewardship programs in Australia, for packaging and batteries. She has published widely on product stewardship and sustainable packaging. Her most recent book, *Product Stewardship in Action: the Business Case for Lifecycle Thinking*, was published in 2016.

Vanessa A. Masterson is a researcher and theme leader of the Stewardship and Transformative Futures theme at the Stockholm Resilience Centre. She is a visiting researcher at the Department of Geography at Exeter University and an honorary associate at Rhodes University. She holds a PhD (Stockholm University, 2017) in Sustainability Science through which she explored the role of sense of place and culturally mediated rituals in stewardship of small-scale agricultural landscapes and community conservation interventions in South Africa. Masterson's current research draws attention to the cultural values of nature, which play a critical role in people's wellbeing in the context of rural–urban migration in the developing world. She is increasingly interested in exploring how the meanings that we attach to nature support stewardship of landscapes and our collective ability to cope with rapid change and transformation.

Chris Mounsey worked for several years in theater before an accident and four months of immobility, in which reading was the only possible occupation, led to an academic career. Degrees in Philosophy, Comparative Literature and English from the University of Warwick followed, and a doctorate on Blake founded an interest in the literature of the eighteenth century. Chris now teaches at the University of Winchester and is the author of *Christopher Smart: Clown of God* (2001), *Being the Body of Christ* (2012), and *Sight Correction: Vision and Blindness in Eighteenth-Century Britain* (2019). Chris has also edited *Presenting Gender* (2001), *Queer People* (2007), *The Idea of Disability in the Eighteenth Century* (2014), *Developments in the Histories of Sexualities* (2015), *A Spy on Eliza Haywood* (2021) as well as *The Variable Body in History* (2019), *Reconsidering Extinction* (2021), and *The History and Bioethics of Medical Education* (2021) for this series.

Pramod K. Nayar teaches at the Department of English, the University of Hyderabad, India. His most recent books include *Alzheimer's Disease Memoirs* (Springer 2021), *The Human Rights Graphic Novel* (Routledge 2021), *Essays in Celebrity Culture* (2021), *Indian Travel Writing in the Age of Empire* (2020), and *Ecoprecarity* (Routledge 2019). Forthcoming is a collection of essays on Life Writing from Orient BlackSwan.

Tyra A. Olstad has worked as a park ranger, paleontology technician, cave guide, summit steward, and Assistant Professor of Geography and Environmental Sustainability. In addition to two books – *Zen of the Plains* (2014) and *Canyon, Mountain, Cloud* (2021) – she has published a variety of research articles, creative non-fiction essays, photo essays, and hand-drawn maps in numerous academic and literary journals. She is currently practicing place-based stewardship as a Physical Scientist for the National Park Service in southern Utah.

Bruce Rocheleau is Professor Emeritus at Northern Illinois University. He has authored five books as well as many articles and chapters. He has focused on wildlife conservation for his last two: *Wildlife Politics* (2017) and *Industry First: The Attack on Conservation by Trump's Interior* (2021). He maintains a blog on wildlife conservation issues at https://www.wildlifepolitics.org/.

Nathan F. Sayre is Professor of Geography at the University of California, Berkeley. He is the author of *The Politics of Scale: A History of Rangeland Science* (2017); *Working Wilderness: the Malpai Borderlands Group and the Future of the Western Range* (2005); *Ranching, Urbanization, and Endangered Species in the Southwest: Species of Capital* (2002); and *The New Ranch Handbook: A Guide to Restoring Western Rangelands* (2001). He holds affiliations with the USDA-Agricultural Research Service-Jornada Experimental Range, the USDA's Southwest Climate Hub, UC-Berkeley's Energy Resources Group and Department of Anthropology, the Berkeley Food Institute, and the Range Graduate Group.

Matthew Wynn Sivils is a Dean's Professor at Iowa State University, where he teaches courses on nineteenth-century American literature and the environmental humanities. His recent books include an edition of Harriet Prescott Spofford's 1859 novel, *Sir Rohan's Ghost* (2020) and the collection, *Ecogothic in Nineteenth-Century American Literature* (co-edited with Dawn Keetley, 2017).

Uno Svedin is a retired Professor and is still very active as Senior Researcher at the Stockholm Resilience Centre, Stockholm University. He has been actively involved in sustainability issues since the end of the 1970s in many functions – in Sweden and internationally, e.g., in relation to the EU, the UN and in international NGOs. He has been involved in contributing to the development of the framing of "sustainability science," issues also about "planetary boundaries," as well as cooperating with the other chapter authors of this piece in several articles over the last few years on "sense of place" and stewardship. He has also, in other contexts, been addressing similar socio-ecological and culturally oriented sustainability topics over a considerable period.

Maria Tengö is a Principal Researcher at the Stockholm Resilience Centre (Stockholm University). She applies a social-ecological systems approach to address conservation and sustainable use of biodiversity and ecosystems, environmental governance, and sustainability transformations. She has extensive experience in developing theory and practice for transdisciplinary research, engaging in partnership with societal actors at local, national, international level for co-production of knowledge and action for sustainability. In recent years a focus of her research has focused on the interface between Indigenous, local, and scientific knowledge systems, and conditions for successful collaborations for environmental governance and human rights.

Josh A. Weinstein is Associate Professor of English and a faculty member in the Environmental Studies Program at Virginia Wesleyan University, in Virginia Beach, Virginia. He received his PhD in English from the State University of New York at Buffalo in 2007, and his research and teaching interests include American nature writing and ecopoetry, with an eye toward ethical engagement with texts. He has published on a variety of topics, including humility in ecocriticism, and authors such as Marianne Moore, Susan Cooper, and Gary Snyder. He is founding co-editor of *Green Humanities: A Journal of Ecological Thought in Literature, Philosophy & the Arts* <www.greenhumanities.org>.

Simon West is a Researcher at the Stockholm Resilience Centre (Stockholm University). His research focuses on how people generate, share, and use knowledge within the everyday practices of environmental governance. He currently holds a Mobility Starting Grant from the Swedish Research Council Formas, and his projects include examining the co-production of knowledge in the context of Indigenous land management in Northern Australia, and collaboratively exploring human responses to environmental change in Northern Alaska. He is an Honorary Lecturer at the Fenner School of Environment and Society (Australian National University), and a Visiting Fellow at the Northern Institute (Charles Darwin University).

Introduction
The Contradictory Inflections of Stewardship

Rachel Carnell

The idea of stewardship, human responsibility to tend and till the earth, has been central to human cultures throughout history, as evident in the Judeo-Christian Genesis story and a diverse range of tales from other traditions around the world. Despite such foundational hortatory stories about cultivating the land on which we live, humanity is nevertheless currently destroying the planet with breathtaking speed. This contradiction speaks to a central feature of our human condition: we like to think of ourselves as careful stewards of the earth, but aspects of our "stewardship" have been inherently destructive. Our failures to steward carefully may be enabled in part by words we use to describe our responsibility to the planet and its myriad species. For example, has the word "dominion," usual in English translations of the Genesis story, led us astray? As the primatologist Jane Goodall observes, "When I think of our attitude to animals in Genesis, where man is told that he has 'dominion' over the birds and the fish and the animals and so on – the actual word, I'm told, is not dominion, it's stewardship."[1] However, even if humans understand God's command to Adam and Eve to steward the earth rather than to rule over it, the very idea of "stewardship" is also regularly used to refer to activities that do not involve protecting the earth or its flora and fauna.

The Old English *stigweard* does not mean "keeper of the pig-sties" (*OED*), as was once supposed.[2] However, the evolution in meaning from the imagined etymology of the pig keeper, whose animals rooted in and helped till the soil, to the financial officer who became steward or manager of the financial ledgers of a landed estate, is nevertheless revealing.[3] Tending the soil or stewarding the planet is a different responsibility than tending the account books or maximizing profits. Yet that distinction gradually became blurred as landholdings for the wealthy became increasingly viewed in terms of how much they generated in rents, produced from tenants farming their acres on the larger estate. When the regency novelist Jane Austen introduces Fitzwilliam Darcy in *Pride and Prejudice* (1813), she describes his estates in the terminology of her day, in which land is converted to its monetized production value of annualized income.

DOI: 10.4324/9781003219064-1

We learn of his having "ten thousand a year" before we are told of the "large estate in Derbyshire" that produces that income.[4] Austen does not critique Darcy for earning his income from his landholding; in fact, she implicitly praises him for being a careful steward of his estates – attentive to his tenants, his servants, and his family. Other landowners in her novels, however, do face her satirical critique for their reckless attitudes to their land, the ancient trees that grow on it, and the tenants who farm it.[5]

In American history, research for the 1619 Project (the date slaves were first brought to the United States) further reminds us of the potential paradoxes of stewardship: tending large estates run with slave labor was not only unethical but also contributed to the ecological destruction of the soil.[6] These historical facts contradict many plantation owners' rosy self-perception as careful stewards of their land as well as the often-rosy depiction of plantations as "bucolic villages" in Southern romantic literature.[7]

Much scientific research on stewardship today (in fields from ecology to geography, landscape architecture, systems theory, and consumer studies) offers insights that should help address the ecological challenges facing the planet. Simultaneous scholarship on cultural history, religious studies, and literature reminds us that the damage done to the planet has often been carried out in the name of tending the land. As we progress further into the Anthropocene, attitudes today toward stewardship, and the language used to describe them, will be essential to saving the planet. Nevertheless, the academic disciplines are not talking to each other in a language that all can understand.

As professors of eighteenth-century literature, Chris and I were drawn to the category of stewardship as a way of understanding eighteenth-century attitudes toward land and the production of wealth. In that process, we discovered that monetary expectations for estate "rents" often put the term "stewardship" at odds with a different goal of tending the land for the long-term benefit of the soil and the flora and fauna living on it.[8] Meanwhile today, in the world of investment, stewardship usually refers to stewarding funds rather than flora or fauna. For left-leaning Christians, stewardship often refers to an ethical responsibility to the earth, while for some right-leaning Christians, tending the planet may seem pagan or pantheistic, i.e. not the central focus of those Christians and their need to save their individual souls.[9] Even within contemporary academic disciplines of geography, urban studies, oceans research, and green business practice, the term "stewardship," while used to refer to taking care of the earth and its species, is used with many different inflections.

In urban studies, stewardship refers not just to the planet and its flora and fauna, but to "the present generation's inherited endowment of culture, knowledge, and resources."[10] In product stewardship, approaches vary along a spectrum from producer to consumer responsibility, depending on whether the stewardship practices stem from government regulation to industry self-regulation. In contemporary literature, certain dystopian novelists imagine humans in competition with post-human or alien lifeforms. Meanwhile, a group of researchers into resilience have identified four different ways that organizations and citizens often view stewardship: as an ethic, a motivation, an action, or an outcome. These different views and approaches, moreover, are reflected in the different types of agencies that are attempting to effect change: government-led, civic-led, and corporate-led. Thus, even scholars with similar well-intentioned ideals each bring slightly different viewpoints and emphases to the topic of stewardship.

Our goal in this collection of essays is to bring together research on stewardship from across the sciences, social sciences, and humanities to shed light on when researchers in different fields may be speaking the same language and when acts of translation may be required. We believe these different research fields will be well served by a broader understanding of how others in parallel and tangential areas view and discuss the challenges inherent in achieving the ideal of stewarding the earth.

This volume of essays offers a sampling of research across the academic disciplines, showcasing the different ways that stewardship is being studied. Some researchers take an historical approach, investigating how cultural ideas of "good stewardship" have sometimes allowed humans to ravage the planet; other scholars consider contemporary mindsets – e.g., how we develop a "sense of place" – that may or may not inspire humans to be the best stewards of the whole planet. Other scholars explore literature – including works from the eighteenth and nineteenth centuries as well as dystopian fiction that envisions dark versions of the future that may face humanity if we do not soon succeed in stewarding the planet and its resources. Some scholars examine the scientific progress that has been made in the arena of conservation, for example through the U.S. Endangered Species Act of 1973, while others delineate the challenges facing humans attempting to be good stewards of rangelands, national parks, and the oceans. In the field of "product stewardship," scholars have investigated the intended and unintended consequences of various initiatives to shift the financial burden of waste and packaging from consumers to manufacturers and importers. This research is particularly relevant to contemporary debates in communities where producer-responsibility recycling programs are currently being deliberated.[11]

Although the idea of stewardship has been central to how humans have viewed their responsibility toward the planet, there are no other books that have addressed stewardship as a discourse that holds both promise and peril. The essays in *Earth Stewardship: Linking Ecology and Ethics in Theory and Practice* (2015) take the concept of stewardship as a science and a strategy for change rather than a discourse that has sometimes enabled complacency. Johnny Wei-Bing Lin's *The Nature of Environmental Stewardship* (2016) is written to bridge conversations between evangelical Christians and other, more scientifically minded persons interested in saving the planet, but it is not designed to speak across the different academic disciplines. The present volume is intended to begin interdisciplinary conversations, and especially those between science, social science, and the humanities.

As part of the series *Advances in the History of Bioethics*, this volume takes seriously Ruth Schwartz Cowan's observation that "many bioethical puzzles need to be resolved with historians' tools, because the puzzles are unprecedented and the usual tools – philosophical and theological – are not doing the job."[12] This collection benefits from the perspective of history in that several of the essays demonstrate how the idea of stewardship has sometimes incorporated destructive tendencies toward the earth and its species. Our approach is also consistent with the perspective of the seventeenth-century philosopher John Locke, who observed the obstacles to communication that ensue "when any word does not excite in the hearer the same idea which it stands for in the mind of the listener."[13] Our goal with this collection is to help foster a more informed and interdisciplinary understanding of the different approaches currently being taken in the name of stewardship. With that goal in mind, we have divided the volume into four sections of essays that most closely dialogue with one another; however, we hope readers will explore essays written from different and unfamiliar perspectives and find multiple points of connection between them.

Overview of the Volume

The essays in this collection are divided into four parts. The first part reflects on historical views of stewardship, from Robinson Crusoe's attitude toward the island on which he was stranded, to religious attitudes toward land stewardship in nineteenth-century American literature, to the paradoxical effects of our modern-day "sense of place" toward parks and wilderness areas. Essays in the second part examine dystopian visions of human stewardship in fiction from the nineteenth century to the present day. The third part explores the promises and challenges of stewardship practices in real-world situations – from the success of the U.S. Endangered Species Act to the effects of global environmental change on oceans and islands, to the common misperceptions about rangeland that impede fact-based approaches to its long-term management. The essays in the fourth part step back and envision the future: first, in considering the most effective ways to be mindful consumers; second, in examining the paradoxes of product stewardship; third, in applying systems theory to theorize how best to protect the future of humanity on this planet; fourth, in recommending how we might incorporate the core concepts of care, knowledge, and agency to "nurture pluralistic and critically reflexive approaches to stewardship."[14]

Human Self-Perception and Misperception

The volume begins with Tyra Olstad's essay on of "sense of place," an expression of human attachment to special places – an affective attachment that, as Olstad demonstrates, does not automatically translate into an ethic of caring. Olstad explores why sense of place translates into an ethic of stewardship in some cases but not in others and how we can learn from existing assumptions and ideals to build stronger senses of place and enhance successful place-based stewardship efforts. Our sense of ourselves in nature, of course, is not just dependent on national park brochures that encourage us to fall in love with a special place in the natural world. Our sense of ourselves is also shaped by the stories and novels we have read or heard about human nature throughout cultural history. Daniel Defoe's *Robinson Crusoe*, the 1719 novel about the shipwreck survivor who thrives alone for years on an island, is known for its celebration of economic individualism and labor, yet, as Chris Mounsey demonstrates, the lengthy description that Defoe offers of the effort required for Crusoe to make his first loaf of bread is less focused on his labor per se than on Crusoe's gradual recognition of his Christian responsibilities to the earth. As Mounsey explains,

> the care Crusoe takes in making a loaf of bread, and the context in which he makes it, becomes a metaphor for the relationship between each human reader and the place in which they find themselves. We must all remember, Crusoe reminds us, to take the same care in our relationship with whatever place we find ourselves in the world and in the things we do to get our bread.[15]

Moving to nineteenth- and twentieth-century American literature, Josh Weinstein investigates the lines of intersection and divergence between the religiously motivated and scientifically motivated strains of stewardship through the works of the American writers Susan Cooper, Aldo Leopold, and Gary Snyder. As Weinstein points out,

> A key challenge in achieving practical synergy in addressing environmental challenges between adherents of these distinct, yet convergent viewpoints, may be figuring out how to maintain focus on the common ground of a belief in caring for the natural world, while fostering an openness to a diversity of opinions on the first principles that motivate disparate groups to do so.[16]

Dystopian Visions of Past, Present, and Future

The second part of this collection focuses on dystopian and utopian views of stewardship, beginning with Matthew Wynn Sivils' analysis of the nineteenth-century cultural imagery of the bucolic village so often associated with American slave-owning plantations. Sivils demonstrates that the nineteenth-century texts promoting this bucolic vision in fact also reveal "an environmental grotesque born of a monstrous stewardship."[17] Pramod K. Nayar addresses the complex questions raised by human stewardship of the human genome itself in his analysis of late twentieth- and early twenty-first-century dystopian fiction by Margaret Atwood, Kazuo Ishiguro, and Octavia Butler. As he explains, "a coercive placental economy is visible in the dystopian vision of these writers, where there is potential for miscegenation between humans and alien lifeforms, men, women and surrogate mothers, and humans and 'their' clones respectively." Nayar concludes that "Human stewardship in Octavia Butler, Margaret Atwood and Kazuo Ishiguro is thus envisioned as a violent, hierarchic and exploitative prospect."[18] Jeff Karem explores how Paolo Bacigalupi's recent *Drowned Cities* trilogy goes even further, upending contemporary expectations of human and environmental stewardship by reframing climate change as a corrective response to human excesses. Rather than affirming the power of human agency to save the earth, "Bacigalupi's novels demonstrate that the earth does not need saving." As Karem explains, "His fiction proposes that the earth, as a global system, will protect itself by purging itself of organisms with unsustainable ways of life, which includes the majority of twenty-first century humanity."[19]

Approaches to Contemporary Challenges

The third part of the volume addresses pragmatic approaches that have been successful to improved stewardship. Bruce Rocheleau traces the history and effect of the U.S. Endangered Species Act of 1973, a federal law that appeals to public appreciation for the "intrinsic value" of certain endangered species, a piece of legislation and has helped counter the strong forces of utilitarian and consummatory capitalism in U.S. culture, forces that continue to threaten many species today. Nathan Sayre explains how we need to change our long-held views on stewarding the 40 percent of the earth that is considered "rangeland," a broad category that includes "any landscape that is neither forested, cultivated, buried in ice, built up or paved over." As he points out, viewing such lands as marginal has tended to allow them to be exploited. We thus "require new narratives that elevate rangelands for their beauty and positive values, rather than just their vast extent and putative degradation." Such stewardship must also involve "strategies that engage and strengthen local communities and institutions vis-à-vis outside forces – including scientists – whose ambitions, when not openly predatory, are often still suffused with flawed assumptions and wishful thinking."[20] Sun-Kee Hong concludes this part with an essay reflecting on how traditional cultural knowledge of islanders in the Korean archipelago must be yoked together with contemporary climate stewardship as we seek to preserve both the future of islands and the future of the planet.

Envisioning the Future

The essays in the final part of this collection reflect on the past with an eye to stewarding the future of the planet and its species. Helen Lewis and Nick Florin consider the paradoxes of "product stewardship," i.e., ecologically minded practices intended to reduce the environmental and social impacts of products by allocating more responsibility to producers and importers. While important to sustainability, these practices nevertheless may "help to support or justify growing levels of production and consumption as 'business as usual.'"[21] William M. Bowen recommends an evolutionary systems theoretic perspective to help us recognize "which of the factors in today's large-scale complex systems can be changed through stewardship, which cannot, and what interventions appear most apt to go furthest toward preserving, protecting, and passing on the current inherited endowment for future generations." He concludes that we will need "better direct person-to-person communication, recognition of the interdependence of individuals and groups (including the implications of collective failure), and the creation of cultural conventions, social norms, and institutions conducive to enhanced levels of cooperation."[22]

In the final essay of the volume, Maria Tengö and a group of researchers at the Stockholm Resilience Centre consider the limits of our traditional views of stewardship and suggest how they may be reinvigorated in the Anthropocene through new attitudes toward human–nature reciprocity. They conclude that humans may be able to "promote more relational perspectives on where we ascribe values embedded in world views based on human–nature reciprocity, and the related transformative capacity towards sustainability in everyday life." They also suggest we view the category of stewardship as a "boundary object" – a conceptual tool that "enables collaboration and dialogue between different actors whilst allowing for differences in use and perception."[23] This approach to understanding stewardship as one that straddles the boundaries between different actors and different ways of thinking, sums up the meta-disciplinary goals of this entire collection of essays. In bringing together the different essays in this collection, we hope to encourage precisely this sort of collaboration and dialogue between the different approaches that scholars and activists around the world are taking to preserve the future of the earth and its species.

Notes

1 Cited in David Marchese, "Why Jane Goodall Still Has Hope for Us Humans." *The New York Times*, July 12, 2021.

https://www.nytimes.com/interactive/2021/07/12/magazine/jane-goodall-interview.html?searchResultPosition=1. Goodall here may not be referring to a problem with the translation so much as the inference. My colleague Mark Wirtz, a biblical scholar, looked into this for me and reports that "I believe Jane Goodall is referring to the Hebrew verb in Genesis 1:26 and 28 that is commonly translated as 'rule' or 'have dominion' (*və·yir·dū* in v. 26 and *ū·rə·dū* in v. 28, verbal forms of the root *rada*). As has often been the case, one could assert the idea of a benevolent rule, but this becomes difficult in verse 28 where God first tells humankind to 'subdue' (*və·ḵiḇ·šu·hā* from the root *kavash*) the earth: ' …and God said to them, "Be fruitful and multiply, and fill the earth and *subdue* it; and *have dominion* over the fish of the sea…"' (NRSV, emphasis mine). Considering that *kavash* can be equally translated as 'dominate' as it can be 'subdue', a harsher, even violent context for a so-called dominion continues to emerge. One could, however, infer from the broader creation narrative a theological/ethical interpretation of stewardship or tending to nature, rather than one of reckless exploitation. For example, if God creates humankind in God's own image in Gen. 1:26–27, and if God creates 'the human' (*hā·'ā·ḏām*, also 'the man', from the root *adam*) from 'the earth' (*hā·'ă·ḏā·māh* [the same root] 'the earth', 'soil', 'ground', 'land') in Gen. 2:7, then one can reasonably say that we have a unique relationship with the Creator and creation. If we are to be good, divinely blessed human beings fashioned from the earth itself, then we have a special duty to not harm the environment; rather, our commission is to be ecologically sensitive stewards of the earth" (Mark Wirtz, email to author, October 4, 2021). See also, Jessica Lane, "'Dominion Over All the Earth': Exploitation or Stewardship?" *Christian Science Monitor.* April 21, 2011. https://www.csmonitor.com/Commentary/A-Christian-Science-Perspective/2011/0421/Dominion-over-all-the-earth-exploitation-or-stewardship. Accessed July 19, 2021.
2. This mistaken origin was apparently widespread enough that in its etymology for the word "steward," the *Oxford English Dictionary* makes clear that "there is no ground for the assumption that *stigweard* originally meant 'keeper of the pig-sties'." See "steward, n." OED Online. Accessed June 2021. Oxford University Press. https://www-oed-com.proxy.ulib.csuohio.edu/view/Entry/190087?rskey=xfjef0&result=1.
3. For a discussion of the eighteenth-century practice of allowing pigs into apple orchards to improve the soil, see Chapter 4 of Chris Mounsey, *Christopher Smart: Clown of God* (Lewisburg, PA: Bucknell University Press, 2003). See also, New Terra Farm, "Raising Pigs in the Garden," https://www.new-terra-natural-food.com/pigs-in-the-garden.html. Accessed July 21, 2021.
4. Jane Austen, *Pride and Prejudice* (Oxford: Oxford World's Classics, 2004), 6.
5. Austen praises Darcy for his attentive attitude toward the servants and tenants through the voice of his housekeeper, Mrs. Reynolds. She also mocks Darcy's aunt, Lady Catherine de Bourgh, for her officious but less empathetic attitude toward the cottagers on her estates: "whenever any of the cottagers were disposed to be too quarrelsome, discontented or too poor, she sallied forth into the village to settle their differences, silence their complaints, and scold them into harmony and plenty" (*Pride and Prejudice*, 130). Austen also satirizes James Rushworth in *Mansfield Park* for his readiness to cut down an old avenue of oaks in the name of making "improvements" to the estate. See Jane Austen, *Mansfield Park* (New York NY: Norton Critical Editions, 1998), 40–41. While she does not directly critique the concentration of wealth in the hands of the upper gentry or the practice of viewing estates in terms of their expected monetary production, Austen does suggest that owners of large estates have a responsibility to steward them carefully. It is "business with his

steward" that brings Darcy back to his estate sooner than expected, ultimately propelling the marriage plot forward as Elizabeth revises her initial view of him when she recognizes, "As a brother, landlord, a master, how many people's happiness were in his guardianship" (*Pride and Prejudice*, 194, 189).

6 Matthew Desmond, "In Order to Understand the Brutality of American Capitalism, You Have to Start on the Plantation," *The New York Times*, August 14, 2019. https://www.nytimes.com/interactive/2019/08/14/magazine/slavery-capitalism.html. Accessed July 20, 2021. This sort of analysis was already evident in work by scholars such as Paul Outka, who writes that slavery was "coextensive with white stewardship of a pastoral landscape." See Outka, *Race and Nature*, 103, cited in Matthew Sivils, "Monstrous Stewardship and the Plantation in Charles Chesnutt's 'The Goophered Grapevine,'" below, 76, 86.

7 See Matthew Sivils' analysis of the paradoxical image in American literature of the "outwardly bucolic slave plantation" an image that belies the reality of it as a "site of a horrifying amalgam of racial oppression and environmental exploitation," below, 75.

8 This collection of essays owes its point of origin to a letter in the Blenheim manuscript collection at the British Library, a letter that I happened across by chance when researching a very different topic. That letter referred to a widow taking a lawsuit to claim the promised value of an estate granted her in a settlement. Her complaint was that the monetary amount produced by the estate in question did not match the amount she had understood the per annum value to represent. (Unfortunately, because the letter was not related to my actual research project, I did not transcribe it or make note of the shelfmark but merely discussed it with my coeditor as an interesting origin point for the idea of extractive capitalism.) Since that other research project was supported by a Faculty Scholarship Initiative Grant from Cleveland State University's Office of Research in 2019, I once again express my gratitude for that institutional support, which has resulted in this second book project, as an unexpected spinoff from the first book that grant supported.

9 See Bernard Daley Zaleha and Andrew Szasz, "Why Conservative Christians Don't Believe in Climate Change," *Bulletin of the Atomic Scientists* 71, no. 5 (2015), 19–30.

10 See William M. Bowen, "An Evolutionary Systems Theoretic Perspective on Global Stewardship," below, 212.

11 See also Winston Choi-Schagrin, "Maine Will Make Companies Pay for Recycling. Here's How it Works." *The New York Times*, July 21, 2021, online edition. Accessed July 28, 2021.

12 Ruth Schwartz Cowan, *Heredity and Hope: The Case for Genetic Screening* (Cambridge MA: Harvard University Press, 2008), 9.

13 John Locke, *An Essay Concerning Human Understanding*, 20th edition, 2 volumes (London: T. Longman, 1796), vol. 2, 7.

14 Maria Tengö et al, "Stewardship in the Anthropocene: Meanings, Tensions, Futures," below, 236.

15 "'I was under no Necessity of seeking my Bread'": *Robinson Crusoe* and the Stewardship of Resources in Eighteenth-Century England," below, 35.

16 See Josh Weinstein, "Stewardship in American Literature: Promise and Paradox in the New World," below, 61.

17 See Sivils, "Monstrous Stewardship," below 76.

18 See Pramod K. Nayar, "Human Stewardship and 'Reproductive Futurism' in Dystopian Fiction," below, 100.

19 See Jeff Karem, "Climate Change and Apocalyptic Literature: Post-Human Stewardship in Paolo Bacigalupi's *Drowned Cities* Trilogy," below, 108.

20 See below, in Nathan Sayre, "Stewardship of Rangelands in the 21st Century: Managing complexity from the margins," below, 186.
21 Lewis and Florin, "Product Stewardship," below, 239.
22 See William Bowen, "An Evolutionary Systems Theoretic Perspective," below, 231.
23 See Maria Tengo et al, "Stewardship in the Anthropocene," below, 236.

Bibliography

Austen, Jane. *Pride and Prejudice*. Oxford: Oxford World's Classics, 2004.
Austen, Jane. *Mansfield Park*. New York: Norton Critical Editions, 1998.
Choi-Schagrin, Winston. "Maine Will Make Companies Pay for Recycling. Here's How it Works." *The New York Times*, July 21, 2021. https://www.nytimes.com/2021/07/21/climate/maine-recycling-law-EPR.html?searchResultPosition=1.
Cowan, Ruth Schwartz. *Heredity and Hope: The Case for Genetic Screening*. Cambridge, MA: Harvard University Press, 2008.
Desmond, Matthew. "In Order to Understand the Brutality of American Capitalism, You Have to Start on the Plantation." *New York Times*, August 14, 2019. https://www.nytimes.com/interactive/2019/08/14/magazine/slavery-capitalism.html.
Lane, Jessica. "'Dominion Over All the Earth': Exploitation or Stewardship?" *Christian Science Monitor*. April 21, 2011. https://www.csmonitor.com/Commentary/A-Christian-Science-Perspective/2011/0421/Dominion-over-all-the-earth-exploitation-or-stewardship.
Locke, John. *An Essay Concerning Human Understanding*, 20th edition, 2 volumes. London: T. Longman, 1796.
Marchese, David. "Why Jane Goodall Still Has Hope for Us Humans." *The New York Times*, July 12, 2021. https://www.nytimes.com/interactive/2021/07/12/magazine/jane-goodall-interview.html?searchResultPosition=1.
Mounsey, Chris. *Christopher Smart: Clown of God*. Lewisburg: Bucknell University Press, 2001.
New Terra Farm. "Raising Pigs in the Garden." https://www.new-terra-natural-food.com/pigs-in-the-garden.html.
Wirtz, Mark, email to Rachel Carnell, 4 October 2021.
Zaleha, Bernard Daley and Andrew Szasz. "Why Conservative Christians Don't Believe in Climate Change." *Bulletin of the Atomic Scientists* 71, no. 5 (2015): 19–30.

Part I
Human Self-Perception and Misperception

1 Stewardship and Sense of Place
Assumptions and Ideals

Tyra A. Olstad

The Premise

"[T]his is not hard to understand," writes author and activist Terry Tempest Williams in simplifying the relationship between sense of place and place-based stewardship: "falling in love with a place, being in love with a place, wanting to care for a place and see it remain intact as a wild piece of the planet."[1] Of course, this makes sense. If and when an individual develops a relationship with a location – be it a national park or a city street, a historic landmark or a local brewpub, or their own backyard – they will want to maintain if not improve it, so that they can continue to enjoy it, so that others can also experience it, because it brings emotional fulfilment, and/or because it just feels like the right thing to do, to take care of that which we love.

A century and a half ago, when American conservationists were attempting to convince the United States Congress to do something unprecedented – to set aside an obscure corner of a distant territory as a national park, to be managed in the public interest – they invited legislators and citizens to experience the landscape vicariously, using grand paintings, dramatic photographs, and compelling written accounts to introduce Americans to previously unknown marvels and make obvious the need to preserve the place in perpetuity. A half-century later, the director of the newly created National Park Service (NPS) launched a public outreach campaign, collaborating with railroad companies and automobile organizations to promote travel to and, by extension, generate support for the parks. Fast-forward to the mid-twentieth century, when increasing industrial development threatened the integrity of parks and protected areas: non-governmental organizations fought proposals for dams and mines by, again, using images, texts, advertisements, and, especially, tourism, to acquaint citizens with their public lands and inspire a demand for preservation. *This Is Dinosaur* – a book of essays and photographs published by the Sierra Club revealed "Echo Park Country and Its Magic Rivers" to the American people, assuming that once people knew about Dinosaur National Monument and its "magic," they would want it protected and well cared for.

That assumption – with all of its promise and peril – lives on today, in an era when tourism bureaus, businesses, local and national governmental agencies, non-profit organizations, scientists, artists, and changemakers can and do use documentary films, social media campaigns, advertisements, and traditional media to convince people to visit places and/or learn place-specific "facts," believing that once a person learns about a location, they will automatically develop a relationship with it and want it to remain intact. Even more so, an ethic of "stewardship" is emerging, in which citizens themselves become caretakers of parks and protected areas, actively helping protect places, instead of merely offering political and economic support. From Yellowstone to Yosemite, Alaska Geographic to the Adirondack Mountain Club, agencies and organizations seek to facilitate and engage an ethic of place-based stewardship.

But is it such a simple progression from "Explore" to "Learn" to, ultimately, "Protect," as the "Junior Ranger" motto of the NPS reads? Are visions of place-based protection and stewardship as seamless and uniform as individuals and organizations might like to believe?

Of course not. Questions then must shift: what are our place-based assumptions and ideals, and what do we need to rethink? Why does sense of place translate into an ethic of stewardship in some cases and not in others? How can we build stronger senses of place and enhance successful place-based stewardship efforts? In other words, why and how might people fall in love with a place, and why and how might they care for it?

Sense of Place

In its broadest interpretation, the term *sense of place* encompasses the physical, ideological, and emotional relationships humans have with geographic locations. It is more of an ongoing process than an end state, as people continuously interact with and layer perceptions on the world, adding knowledge and value to otherwise undifferentiated space[2] and continuously renegotiating personal and societal "meanings, beliefs, symbols, values, and feelings" ascribed to localities.[3] A basic "sense of place" can form through literal sensation – first-hand experience of a geographic location, in all its sensual richness – and/or through second-hand exposure, via personal accounts or written and/or audiovisual representations in books, magazine articles, websites, and films.[4] In both cases, "sense of place" goes beyond sensation to engage an element of cognition, getting a "feel" for the landscape or feature – an idea of what it is and what it means or represents.

Beyond the basic awareness and conceptualization, there are several dimensions and nuances to people–place relationships, ranging from "place meaning" and "place dependence" to "place attachment" and deeply felt "place identity."

"Sense of place" is about the *place* – the where and what. That distinguishes it from *place meaning* – "the descriptive, symbolic meaning that people ascribe to a place."[5] Meanings exist separately from geography and are constructed by and layered upon places via psychological, sociocultural, and sociopolitical processes. Take Yellowstone National Park, for example: its place meaning arises from its wildness and national parkness, with all the bureaucratic, cultural, and symbolic weight that entails; meanwhile, its sense of place centers on its Yellowstone-ness, with its geysers and fumaroles, waterfalls and wolves. The biogeophysical landscape exists on its own, but the *meaning* of that landscape emerges from continuous negotiations and re-negotiations within and between individuals and societies.

Place attachment – "emotional bonds between an individual and a geographic locale, or how strongly a person is connected to a place"[6] – is much more personal. Affective emotional bonds form or do not form based on individual preferences and experiences. Some people like mountains, others like seashores; some prefer quiet rural communities, others crave the exciting urban vibe. Because individuals plant geographic memories like mental flags in places where we live and visit, we all tend to grow more attached to areas where we spend more time – childhood homes, long-time neighborhoods – and/or locations where we experienced meaningful personal events. The longer a person spends in a place, and/or the longer their family or nation has been there, the more it is filled with individual and collective memories and significance. That said, the development of place attachment is not wholly individualistic. Take Yellowstone, again: between its national park status and its mountain-rich landscape, tourists are culturally and aesthetically predisposed to appreciate it more than, say, a nondescript swath of shortgrass prairie a hundred miles east. Designation as a park or protected area indicates and reinforces collective appreciation for specific locations. As Masterson et al. note, "place meanings and attachment are subjective, but they vary systematically."[7]

Place dependence – "the potential of a particular place to satisfy the needs and goals of an individual"[8] – provides enjoyment and fulfillment, but lacks the emotional heft of place attachment. A hiker might appreciate a mountain simply because it has a trail to the top, as opposed to loving the peak itself for its rocks, lichens, moods, and views; a birdwatcher might appreciate a wetland for the opportunity to check species off a list, as opposed to its specific congregation of water, moss, sounds, and scents. If the mountain or wetland is threatened by ecological and/or sociopolitical change, the hiker or birdwatcher might be happy enough to find crags to climb or birds to watch elsewhere. By contrast, *place identity* is an individual or group-level sense of deep belonging, in which geography is seen as a critical part of one's autobiography. If a hiker spent years dreaming of and earning the badge of "Forty-Sixer" – an honor for those who climb all forty-six peaks above four thousand feet in elevation in New York State's Adirondack Mountains – then they might feel that their personal biography is inextricably linked with those specific high points. If a birdwatcher has been returning to the shores of the Adirondacks' Lake Colden for decades to listen for warblers, loons, and the occasional Bicknell's thrush, they might feel that the birdsongs, morning mist, and reflection of Mt. Colden in the calm waters of its namesake lake add up to the most beautiful place on Earth, irreplaceable.

As Ed Abbey wrote of his beloved Arches National Monument: "This is the most beautiful place on earth. Every man, every woman, carries in heart and mind the image of the ideal place, the right place, the one true home, known or unknown, actual or visionary."[9] When individuals come to see a place as ideal, right, or home, our very senses of self can become linked to the integrity of the landscape. To prevent damage to beloved peaks, for example, Forty-Sixers work to maintain trails, monitor wildlife, and/or instruct others on how to recreate responsibly in the Adirondacks.[10] Birdwatchers might argue against new rules and regulations banning campfires and removing lean-tos at Lake Colden. Should a well-known and well-loved arch at Arches crumble – Wall Arch, say, in 2008 – an individual who had felt a deep sense of kinship with the sandstone span may experience an acute feeling of loss and grief.

Sense of place, in all its dimensions, is a universal human experience. Because we have all learned about and/or visited previously unknown spaces, layered meanings on landscapes, and developed feelings of place attachment, if not identity, it is easy to fall into a form of confirmation bias and make several understandable but precarious assumptions: first, that those who experience the same place, either in person or through second-hand media, will ascribe the same place meanings to it; second, that sense of place, place dependence, and place attachment are synonymous; and, third, that simple exposure to a place automatically builds into place attachment and/or identity.

But sense of place, place attachment, place dependence, and place identity are different types of relationships, which vary by person and geography, and, as Masterson et al. go on to try to warn us, "there can be a range of meanings associated with the same place."[11] It is better to think of these assumptions as ideals, which lead to stewardship only when all of the component factors align.[12]

Stewardship

Invoking another broad interpretation, "environmental stewardship" consists of "the actions taken by individuals, groups or networks of actors, with various motivations and levels of capacity, to protect, care for or responsibly use the environment in pursuit of environmental and/or social outcomes in diverse social-ecological contexts."[13] As with sense of place, it is a process: "the active shaping of pathways of social and ecological change for the benefit of ecosystems and society."[14] Especially when considering it in relation to socio-ecological sustainability – efforts to ensure that societies and ecosystems can continue to function and flourish for generations – there is a forward-thinking, altruistic dimension to it. Worrell and Appleby define stewardship as "the responsible use (including conservation) of natural resources in a way that takes full and balanced account of the interests of society, future generations, and other species, as well as of private needs, and accepts significant answerability to society,"[15] while Hernandez emphasizes "the extent to which an individual willingly subjugates his or her personal interests to act in protection of others' long-term welfare."[16]

On a societal level, citizens may engage in a general type of global stewardship by voting for and otherwise supporting political and bureaucratic leaders who work to minimize environmental damage and maximize protection of natural systems and human health. As consumers, they may also choose what products and services to use based on their impacts, economically "voting" to support businesses that minimize damage and maximize protection. Beyond the ballot box and pocketbook, individuals and groups may engage in place-specific stewardship by directly contributing their own time and energy to environmental protection efforts, volunteering for boots-on-the-ground practices such as trail maintenance, trash clean-ups, ecological monitoring, citizen science, and public outreach and education. While some of these efforts are deliberately arranged through grassroots "Friends" groups and/or agency or organizational Volunteer Coordinators, others are encouraged as part of broader cultural shifts, such as adoption of "Leave No Trace" principles for responsible outdoor recreation.

To explain what motivates people to contribute to environmental stewardship efforts, researchers focus not just political or economic factors, but also on social and, especially, psychological dimensions: "values, cognitions, and perceptions of human relationships with nature."[17] For example, Clary et al. developed a "Volunteer Functions Inventory" that measures volunteer motivation in terms of values, understanding, social relationships, protective instincts, career goals, and desire for personal enhancement[18]; Bramston, Petty, and Zammit's "Environmental Stewardship Motivation Scale" adds a "biospheric factor," part of people's larger desire "to do something worthwhile."[19] According to Lee and Hancock, the "most common motivation identified for individuals involved in volunteer stewardship groups is a desire to protect and preserve the natural environment they appreciate and care about."[20]

Sense of place adds geographic specificity and weight to these general perceptions, cognitions, and values. Because sense of place can help develop "ethic of respect" for locations and their broader meanings, it "tends to foster stewardship" for the places themselves and associated resources.[21] At a deeper level, place attachment and place identity afford a "sense of belonging and connectedness to a specific ecological context."[22] This feeling of connectedness can be a strong motivating factor for environmental stewardship.[23] Moreover, there is always a geographic context to stewardship actions – stewardship actions take place in a place. Or, as Masterson et al. write: "stewardship practices and knowledge are embedded in place."[24]

But, as with not-always-accurate assumptions about sense of place, it is all too easy to presuppose that place attachment, dependence, and/or identity translates into an ethic of care. It is a common stewardship credo: "'[t]he more people you connect to [a place] – that you bring to [the place] just to have basic recreation, to educate about [the place], it's going to make them care about it.'"[25] Explore, Learn, Protect: one, two, three. This assumption even infiltrates research: as Masterson et al. observe, "A great deal of the sense of place literature implicitly assumes (or explicitly asserts) that greater place attachment leads to pro-environmental behavior."[26] Assumptions and assertions can fall through or even backfire, though. Any individuals or groups that are hinging their stewardship hopes on cultivating a sense of place need to recognize the potential weaknesses and downsides to place-based initiatives.

Sense of Place ≠ Place Attachment

Just as it may be easy to fall in love with a place, it is easy to *not* fall in love with a place. Simply experiencing an area is not enough; an individual needs to have positive personal ties to, interest in, preference for, and/or compatibility with a location in order to develop a meaningful relationship with it.[27] A person may feel no nostalgic warmth for the place they grew up and/or spend their adult life in a city they hate. A visitor may spend days at a park or historic site and, despite the best attempts by professional guides to facilitate opportunities for them to develop emotional and intellectual connections to the place (the credo of "Interpretation"), fail to connect it if it does not align with their preferences and values.

Beyond our personal biases, we are all culturally conditioned to attach to some places more readily than others. In terms of natural landscapes, we tend to ignore and/or demean "boring" plains and prairies and dismal swamps, preferring rugged mountains, sandy beaches, bucolic river valleys, and lush forests.[28] We are more willing to designate subjectively scenic locations as parks or protected areas, which, in turn, prioritizes these places in the public sphere, drawing more attention, funding, support, and – circling round – appreciation for the scenery. In terms of human-built landscapes, we shun rusty industrial sites and decrepit neighborhoods, preferring tidy suburbs and shiny city centers. Few people will claim to love a feedlot, brownfield, or trash-choked canal; place attachment is unlikely to be a motivating factor in their remediation and/or protection, even if they are the locations that need the most tending-to.

Place Attachment ≠ Positive Action

It may be easy, in turn, to understand how a person may fall in love with a place and want to see it protected, but that does not mean that the person will participate in the protection, much less that the protective efforts are actually beneficial for the place and larger socio-ecological systems. As Chapin and Knapp explain, sense of place does not always give rise to stewardship because: "attitudes (cognitions, emotions, and intentions) may not lead to actions," "some actions may not promote sustainability," and "different place identities that develop in the same place may lead to different, and sometimes conflicting, stewardship goals."[29]

Stewardship entails active change to a place – re-routing of a trail, rethinking management policies, forgoing lawn care in favor of a pollinator-friendly meadow-like setting. Changes, in turn, affect not only the biogeophysical landscape, but people's senses of place. If proposed changes threaten to weaken or undermine an individual's or group's place meanings, attachment, and/or, especially, place identity, those people may fight to prevent the changes, even if, in the long term, they would be the most ecologically and/or socioeconomically sound.[30] For example, every time officials announce management changes – such as removal of high-elevation lean-tos (in the 1970s), group size limits and a campfire ban (1999), and removal of a small dam (2014) – to a popular Wilderness Area in New York's Adirondack State Park, public outcries ensue; even though the changes help mitigate environmental damage and allow for environmental restoration, they also threaten citizens' cherished memories of and expectations for place-specific wilderness experiences.[31] Similarly, Devine-Wright and Howes make the case that instances of "NIMBY"-ism – "Not In My Back Yard" protests, often decried as selfish, exclusionary, and unjust – can actually be traced back to "place protective behavioral responses," in which people want to reject changes to "a location where the project was interpreted to threaten rather than enhance the character of a place."[32]

Chapin and Knapp go a step farther, noting that "sense of place can motivate parochialism and exclusionary practices, as seen in NIMBY attitudes and gated communities"; such reactions "can amplify economic and political disparities," rather than lead to improved conditions.[33] For example, Bonaiuto et al. discuss how "large-scale environmental transformations, like the institution of natural protected areas, can affect people's identity and affective relations with places" and invoke "strong group and 'territorial'" reactions.[34] Their research into attitudes toward

Italian national parks found that local residents "exhibited higher identity, higher attachment and more negative attitudes toward the park (specific and general) than the 'non-locals,'" feeling that the parks "represent something [or some place] that 'others' are trying to take away from 'us.'"[35] Similar resentment festers in the United States, where a vocal minority recently succeeded in removing federal protections for lands previously included in Grand-Staircase Escalante and Bears Ears National Monuments, "taking power away from 'very distant bureaucrats'" and, supposedly, returning it to local residents.[36]

While it would seem like establishment of a park and/or mitigation of environmental damage would equate to better long-term environmental stewardship, the reality is more complicated. Resistance to change can be chalked up to place-based "psychological ownership" – a feeling of possession that "imbues individuals with the internal drive to protect that which is psychologically owned,"[37] which can provoke feelings of threat and defensiveness, as opposed to care. But it may also be that local residents, and/or those with deep emotional attachment to protected places, have deeper understandings of them and all their socio-ecological systems, so are better equipped to care for it than outsiders imposing their non-place-specific and perhaps inappropriate ideals. Even if stewardship in its truest sense "introduces important wider obligations to the wider public, to future generations, and to other species or the natural world," as Worrell and Appleby recognize, "This in turn raises the problem that the diverse groups to whom a steward is responsible will almost inevitably have a range of interests, some of which conflict."[38]

As with all geographic phenomena, scale matters, in terms of both positive action and resistance to change. Individuals may care about environmental issues such as climate change on a global scale, but not want to see wind turbines mar their favorite landscape. Vice versa, a group might work hard to maintain a community garden, but not engage in nation-wide efforts to conserve agrobiodiversity. Although, in general, "place attachment predicts place-specific pro-environmental behavior such as... donation of time and effort in nature refuges...[and] pro-environmental behavior not related to a specific place such as supporting environmental organizations and carpooling,"[39] the mantra "Think Global, Act Local" cannot encompass the complexity of local senses of place (and global values, for that matter) that people may or may not want to act upon.[40]

Conflicting Place Meanings

At heart, sense of place does not translate neatly into stewardship because the same biogeophysical location can mean different things to different people. Take Yellowstone again: is it a bunch of mountains, moose, and geologic oddities? Some pretty scenery? A rare example of a near-intact wild ecosystem? A baseline case for ecological comparison? A sanctuary for wolves and bears? A breeding ground for dangerous predators? Somewhere to go to hike? A destination for a family vacation? A mob of tourists? A symbol of human restraint? America's "best idea"? Land locked up by a federal government? A sell-out to concessionaires? A postcard? An idea? All the same place. When people cannot agree on what the place is and symbolizes, it is challenging to agree on what it *should be*, much less what would it mean to be an effective steward of it. Even Aldo Leopold's seemingly solid criteria for any stewardship actions – "A thing is right when it tends to preserve the integrity, stability, and beauty of the biotic community. It is wrong when it tends otherwise"[41] – offers little concrete guidance. "Integrity" is hard to come by in an era when nearly all Earth places and processes have been modified by man, "stability" is questionable, and "beauty" subjective.

"When different people derive different symbolic meanings from the same place, this can lead to different attitudes, intentions, and actions... despite shared appreciation for the same biophysical features," Chapin and Knapp warn, adding that "different place identities that develop in the same place may lead to different, and sometimes conflicting, stewardship goals."[42] It gets even more complicated when elements of place dependence, attachment, and identity are added in. As Masterson et al. note, "people who are all strongly attached to a place are not necessarily attached to the same thing because one setting can embody many different sets of meanings, each emphasized by different actors."[43] Even when people attempt to care for a place that they love, their ideas of "love" and what the place is and should be often differ, meaning they will care for it differently, and want to see different outcomes for their stewardship efforts. Masterson et al. again: "Debates about the future of a place are thus rarely between people who are attached vs. people who are not, but are rather between holders of different or even oppositional meanings."[44]

Factions can form, especially between perceived "insiders" and "outsiders." As Bonaiuto et al. found, short-term visitors to national parks often have very different understandings about the parks' meanings than do local residents; those coming from different places with different lengths and types of exposure will develop different ideas of "stewardship" for the parks and/or surrounding communities.[45] Chapin and Knapp agree: "[L]ong-term residents may value a place for its capacity to provide them with a livelihood (place dependence) and identity (place identity), whereas newcomers may more often be attracted for its aesthetic or symbolic appeal.[46] Discussing differences between part-time and permanent residents in transitional suburban–rural landscapes, the former more interested in environmental quality and the latter in social relations, Soini, Vaarala, and Poutaa note that "Sense of place is expected to translate into harmony between people and nature, as well as care for the place," but "in an environment with heterogeneous expectations for landscape management," the reality is much more complicated and the variety of place meanings must be taken into account.[47]

When groups or programs attempt to meet place-based objectives by deliberately creating and manipulating place meanings in the public sphere – going back to those nineteenth-century efforts to engender support for a Yellowstone National Park – they are, at a deeper level, capitalizing on and/or seeking to change "people's ethics, morals, values or beliefs."[48] In so doing, they may not accept or acknowledge others' senses of place and views of stewardship. As Cheng, Kruger, and Daniels recognize, groups may even "intentionally manipulate the meanings of places hoping to influence the outcome of natural resource controversies."[49] For example, a coalition of non-profit organizations and for-profit sponsors, such as clothing company The North Face, are currently attempting to engender public support for a ban on energy development in the Arctic National Wildlife Refuge, using documentary films, websites, and magazine articles to encourage American citizens to "Experience the Refuge" – a harsh, wild place that few people will ever visit in person, situated on the far northern edge of the North American continent – then subsequently "advocate... for permanent protection of the Refuge's coastal plain."[50] Images and descriptions of the refuge reveal it as a vast, stunningly beautiful, wholly natural landscape, of spiritual and existential importance to nearby Gwich'in peoples, but do not mention the existence much less the views of local Iñupiat peoples, to whom the refuge is intimately known as their homeland, not just some pretty scenery whose fate should be decided in far-away Washington DC.[51] The very idea of a refuge – lines on a map, reams of management plans – undermines traditional Iñupiat views of land use and ecological connection.[52] Are some place meanings more correct than others? Whose version of stewardship is best? For whom? Who gets to decide?

Because there can be such "substantial variation within and among stakeholder groups in reasons for valuing particular places and therefore the potential for conflicts, as often seen in debates over conservation vs. development among people who value the same place," Chapin and Knapp summarize, "Sense of place is therefore often contested and not a simple panacea for stewardship, as sometimes assumed by environmental advocates."[53]

Lessons and Ideals

Development of a strong sense of place, in all its dimensions, certainly does not guarantee (or even arrive at a consensus definition of) place-based stewardship. But if the assumptions are taken as *possibilities* to negotiate and strive for, rather than guaranteed outcomes, then the process of place creation can give people a framework through which to envision and act upon ideals of stewardship. When individuals or groups build a strong sense of attachment to a place *and* agree and act upon shared place meanings that include broader consideration for the long-term socio-ecological well-being of a location, sense of place can be a powerful motivating force.

This requires careful self-reflection and/or guidance, upon the part of individuals, who need to be willing to hear and acknowledge others' senses of place, and upon the part of groups, which need to be willing to accept that some people will not develop or act upon place attachment and/or share place meanings. Rather than blindly trying to force people to appreciate a place and/or tell them what to do, groups can engage in dialogue that allows for individual expression of place meanings and fosters a deeper sense of place.[54] Individuals, in turn, may find that place-specific feelings and goals can be more powerful motivating forces than abstract values such as "sustainability" and "stewardship."[55]

When these factors align, perhaps the greatest benefit of sense of place is that it can form a positive feedback loop with stewardship. "People–place relations can strengthen protective norms," write Masterson et al., acknowledging that "contested senses of place can be an obstacle for stewardship," then pointing out that "consensus regarding place meanings can contribute to community cohesion and sustainability," further strengthening senses of place and commitment to stewardship.[56] Huq and Burgin agree, concluding that "exchange of views, ideas and knowledge, shared works and networking with like-minded fellow volunteers enhanced the motivational process and their sense of place… This ultimately led to a stronger commitment that drove individual motivation to sustain and contribute further."[57] In other words, when people do help take care a place – *especially* when we invest our own time and sweat in it, and can physically see and take satisfaction in the fruits of our labor – we add another dimension of attachment, if not identity – a little of ourselves left in the landscape, with pride, and, often, a sense of community with other stewards; this can deepen desire to continue stewardship efforts, leading to long-term commitment. "Long-term commitment," in turn, builds more knowledgeable, dedicated volunteers. As Lee and Hancock explain, "This sense of responsibility and stewardship toward natural systems cultivated through participation in restoration activities can result in the creation of a committed constituency of land stewards."[58] Or, as Gary Snyder puts it: "the ecological benefits… of cultivating a sense of place, are that then there will be a people to be the People in the place, when it comes down the line."[59]

Although it is unwise to assume that people will automatically attach to places, take action to protect it, and even agree upon what the significance of the place is and what protection would look like, when individuals and groups acknowledge the complexity of personal and societal senses of place, they have a meaningful opportunity to build upon the foundation of person–place relationships and find motivation for long-term, place-specific, socio-ecological stewardship. To return to Tempest Williams's observation, "this is not hard to understand: falling in love with a place, being in love with a place, wanting to care for a place and see it remain intact as a wild piece of the planet."[60] Across America – and around the world – people can and are helping care for the places they love specifically *because* they love these places.

Notes

1 Terry Tempest Williams, *Red: Passion and Patience in the Desert* (New York: Vintage Books, 2002), 16.
2 Paraphrased from Yi-Fu Tuan, *Space and Place: The Perspective of Experience* (Minneapolis MN: University of Minnesota Press, 1977).
3 Stuart Chapin III and Corrine Knapp, "Sense of Place: A Process for Identifying and Negotiating Potentially Contested Visions of Sustainability," *Environmental Science & Policy* 53 (2015): 40.
4 Alex Kudryavtsev, Marianne Krasny, and Richard Stedman, "The Impact of Environmental Education on Sense of Place among Urban Youth," *Ecosphere* 3, no. 4 (2012).
5 Christopher Raymond, Marketta Kyttä, and Richard Stedman, "Sense of Place, Fast and Slow: The Potential Contributions of Affordance Theory to Sense of Place," *Frontiers in Psychology* 8 (2017): 1.
6 Raymond, Kytta, and Stedman, "Sense of Place, Fast and Slow," 1–2.
7 Vanessa Masterson, Richard C. Stedman, Johan Enqvist, Maria Tengö, Matteo Giusti, Darin Wahl, and Uno Svedin, "The Contribution of Sense of Place to Social-Ecological Systems Research: A Review and Research Agenda," *Ecology and Society* 22, no. 1 (2017), article 49, https://www.ecologyandsociety.org/vol22/iss1/art49/, open access, accessed October 21, 2021.
8 Daniel Williams., Michael E. Patterson, Joseph W. Roggenbuck, and Alan E. Watson, "Beyond the Commodity Metaphor: Examining Emotional and Symbolic Attachment to Place," *Leisure Sciences* 14 (1992): 31.
9 Edward Abbey, *Desert Solitaire: A Season in the Wilderness* (New York NY: Simon & Schuster, 1968), 1.
10 Tyra Olstad, "Visitor Perception, Place Attachment, and Wilderness Management in the Adirondack High Peaks," in *Explorations in PLACE Attachment*, ed. Jeffrey S. Smith (London: Routledge, 2018), 133–148.
11 Masterson et al., "The Contribution of Sense of Place to Social-Ecological Systems Research."
12 Bruce Rocheleau, in another essay this collection, reflects on the affective and contingent factors that likewise determine peoples' willingness to protect charismatic (or less charismatic) endangered species. See "Political Aspects of Stewardship for Wildlife in the U.S.," below, 138–142.
13 Nathan Bennett, Tara S. Whitty, Elena Finkbeiner, Jeremy Pittman, Hannah Bassett, Stefan Gelcich, and Edward H. Allison, "Environmental Stewardship: A Conceptual Review and Analytical Framework" *Environmental Management* 61 (2018): 597.
14 Chapin and Knapp, "Sense of Place," 40.
15 Richard Worrell and Michael C. Appleby, "Stewardship of Natural Resources: Definition, Ethical and Practical Aspects," *Journal of Agricultural and Environmental Ethics* 12 (2000): 269.
16 Morela Hernandez, "Toward an Understanding of the Psychology of Stewardship," *Academy of Management Review* 37, no. 2 (2012): 174.

17 Masterson et al., "The Contribution of Sense of Place to Social-Ecological Systems Research."
18 Clary et al. 1998, described in Paul Bramston, Grace Pretty and Charlie Zammit, "Assessing Environmental Stewardship Motivation," *Environment and Behavior* 43, no. 6 (2011).
19 Bramston, Petty and Zammit, "Assessing," 785.
20 Marty Lee and Paul Hancock, "Restoration and Stewardship Volunteerism," in *Human Dimensions of Ecological Restoration: Integrating Science, Nature, and Culture*, eds. Dave Egan, Evan E. Hjerpe, and Jesse Abrams (Washington, DC: Island Press, 2011), 24.
21 Chapin and Knapp, "Sense of Place," 40.
22 Rafiq Huq and Shelley Burgin, "Eco-social Capital: A Proposal for Exploring the Development of Cohesiveness in Environmental Volunteer Groups," *Third Sector Review* 22, no. 1 (2016): 54.
23 Thomas Measham and Guy Barnett, "Environmental volunteering: motivations, modes and outcomes," *Australian Geographer* 38, no.4 (2008): 537–552.
24 Masterson et al., "The Contribution of Sense of Place to Social-Ecological Systems Research."
25 A volunteer oyster gardener in New York City, quoted in Marianne Krasny, Sarah R. Crestol, Keith G. Tidball, and Richard C. Stedman, "New York City's Oyster Gardeners: Memories and Meanings as Motivations for Volunteer Environmental Stewardship," *Landscape and Urban Planning* 132 (2014): 22.
26 Masterson et al., "The Contribution of Sense of Place to Social-Ecological Systems Research."
27 Kaplan et al., cited in Huq and Burgin "Eco-social Capital," 54.
28 Tyra Olstad, *Zen of the Plains: Experiencing Wild Western Places* (Denton: University of North Texas Press, 2014).
29 Chapin and Knapp "Sense of Place," 41.
30 See Chapin and Knapp "Sense of Place", Masterson et al., "The Contribution of Sense of Place to Social-Ecological Systems Research."
31 Olstad "Visitor Perception."
32 Patrick Devine-Wright and Yuko Howes, "Disruption to Place Attachment and the Protection of Restorative Environments: A Wind Energy Case Study," *Journal of Environmental Psychology* 30 (2010): 273.
33 Chapin and Knapp, "Sense of Place," 39.
34 Marino Bonaiuto, Giuseppe Carrus, Helga Martorella, and Mirilia Bonnes, "Local Identity Processes and Environmental Attitudes in Land Use Changes: The Case of Natural Protected Areas," *Journal of Economic Psychology* 23 (2002), 636.
35 Bonaiuto et al., "Local Identity Processes," 646 and 647.
36 President Donald Trump, quoted in Juliet Eilperin, "A Diminished Monument," *The Washington Post*, January 15, 2019.
37 Hernandez, "Toward an Understanding," 183.
38 Worrell and Appleby, "Stewardship of Natural Resources," 266.
39 Kudryavtsev, Krasny, and Stedman, "The Impact of Environmental Education," 3.
40 See Patrick Devine-Wright, "Think Global, Act Local? The Relevance of Place Attachments and Place Identities in a Climate Changed World," *Global Environmental Change* 23, no. 1 (2013): 61–69.
41 Aldo Leopold, *A Sand County Almanac; and Sketches Here and There* (New York: Oxford University Press, 1949), 211.

42 Chapin and Knapp, "Sense of Place," 41.
43 Masterson et al., "The Contribution of Sense of Place to Social-Ecological Systems Research."
44 Masterson et al., "The Contribution of Sense to Place."
45 Bonaiuto et al., "Local Identity Processes."
46 Chapin and Knapp, "Sense of Place," 42.
47 Katriina Soini, Hanne Vaarala, and Eija Poutaa, "Residents' Sense of Place and Landscape Perceptions at the Rural–Urban Interface," *Landscape and Urban Planning* 104 (2012), 126.
48 Nathan Bennett., Tara S. Whitty, Elena Finkbeiner, Jeremy Pittman, Hannah Bassett, Stefan Gelcich, and Edward H. Allison, "Environmental Stewardship: A Conceptual Review and Analytical Framework," *Environmental Management* 61, no. 4 (2018): 602.
49 Antony Cheng, Linda Kruger, and Steven Daniels, "'Place' as an Integrating Concept in Natural Resource Politics: Propositions for a Social Science Research Agenda," *Society and Natural Resources* 16 (2003): 90.
50 We Are the Arctic, "About Us," accessed May 1, 2020. https://www.wearethearctic.org/about-us.
51 Elizabeth Harball and Nat Herz, "As Oil Drilling Nears in Arctic Refuge, 2 Alaska Villages See Different Futures," *Alaska Public Media*, July 3, 2019.
52 Frank Norris, *Alaska Subsistence: A National Park Service Management History* (Alaska Support Office, National Park Service, 2002).
53 Chapin and Knapp, "Sense of Place," 39.
54 Cheng, Kruger, and Daniels, "'Place' as an Integrating Concept in Natural Resource Politics,'" 95.
55 Olstad, "Visitor Perception," 140.
56 Masterson et al., "The Contribution of Sense of Place to Social-Ecological Systems Research."
57 Huq and Burquin, "Eco-social Capital, 57."
58 Lee and Hannock, "Restoration," 24.
59 Gary Snyder, *The Real Work: Interviews and Talks 1964–1979* (New York: New Directions, 1980), 140.
60 Tempest Williams, *Red*, 16.

Bibliography

Abbey, Edward. *Desert Solitaire: A Season in the Wilderness*. New York: Simon & Schuster, 1968.

Bennett, Nathan J., Tara S. Whitty, Elena Finkbeiner, Jeremy Pittman, Hannah Bassett, Stefan Gelcich, and Edward H. Allison. "Environmental Stewardship: A Conceptual Review and Analytical Framework." *Environmental Management* 61, no. 4 (2018): 597–614.

Bonaiuto, Marino, Giuseppe Carrus, Helga Martorella, and Mirilia Bonnes. "Local Identity Processes and Environmental Attitudes in Land Use Changes: The Case of Natural Protected Areas." *Journal of Economic Psychology* 23 (2002): 631–653.

Bramston, Paul, Grace Pretty, and Charlie Zammit. "Assessing Environmental Stewardship Motivation." *Environment and Behavior* 43, no. 6 (2011): 776–788.

Chapin, F. Stuart III, and Corrine N. Knapp. "Sense of Place: A Process for Identifying and Negotiating Potentially Contested Visions of Sustainability." *Environmental Science & Policy* 53 (2015): 38–46.

Cheng, Antony, Linda Kruger, and Steven Daniels. "'Place' as an Integrating Concept in Natural Resource Politics: Propositions for a Social Science Research Agenda." *Society and Natural Resources* 16 (2003): 87–104.

Devine-Wright, Patrick, and Yuko Howes. "Disruption to Place Attachment and the Protection of Restorative Environments: A Wind Energy Case Study." *Journal of Environmental Psychology* 30 (2010): 271–280.

Devine-Wright, Patrick. "Think Global, Act Local? The Relevance of Place Attachments and Place Identities in a Climate Changed World." *Global Environmental Change* 23 (2013): 61–69.

Eilperin, Juliet. "A Diminished Monument." *The Washington Post*, January 15, 2019. https://www.washingtonpost.com/graphics/2019/national/environment/will-anyone-mine-after-grand-staircase-escalante-reduction-by-trump/.

Harball, Elizabeth and Nat Herz. "As Oil Drilling Nears In Arctic Refuge, 2 Alaska Villages See Different Futures." *Alaska Public Media*, July 3, 2019. https://www.npr.org/2019/07/03/738145007/as-oil-drilling-nears-in-arctic-refuge-2-alaska-villages-see-different-futures.

Hernandez, Morela. "Toward an Understanding of the Psychology of Stewardship." *Academy of Management Review* 37, no. 2 (2012): 172–193.

Huq, Rafiq, and Shelley Burgin. "Eco-social Capital: A Proposal for Exploring the Development of Cohesiveness in Environmental Volunteer Groups." *Third Sector Review* 22, no. 1 (2016): 49–68.

Krasny, Marianne E., Sarah R. Crestol, Keith G. Tidball, and Richard C. Stedman. "New York City's Oyster Gardeners: Memories and Meanings as Motivations for Volunteer Environmental Stewardship." *Landscape and Urban Planning* 132 (2014): 16–25.

Kudryavtsev, Alex, Marianne E. Krasny, and Richard C. Stedman. "The Impact of Environmental Education on Sense of Place among Urban Youth." *Ecosphere* 3, no. 4 (2012): 1–15.

Lee, Marty and Paul Hancock. "Restoration and Stewardship Volunteerism." In *Human Dimensions of Ecological Restoration: Integrating Science, Nature, and Culture*, edited by Dave Egan, Evan E. Hjerpe, and Jesse Abrams, 23–28. Washington, DC: Island Press, 2011.

Leopold, Aldo. *A Sand County Almanac; And Sketches Here and There*. New York: Oxford University Press, 1949.

Masterson, Vanessa, Richard C. Stedman, Johan Enqvist, Maria Tengö, Matteo Giusti, Darin Wahl, and Uno Svedin. "The Contribution of Sense of Place to Social-Ecological Systems Research: A Review and Research Agenda." *Ecology and Society* 22, no. 1 (2017) article 49. https://www.ecologyandsociety.org/vol22/iss1/art49/.

Measham, Thomas and Guy Barnett. "Environmental Volunteering: Motivations, Modes and Outcomes." *Australian Geographer* 39, no. 4 (2008): 537–552.

Norris, Frank. *Alaska Subsistence: A National Park Service Management History*. (Alaska Support Office, National Park Service, 2002), https://www.nps.gov/parkhistory/online_books/norris1/index.htm.

Olstad, Tyra. "Cairns: An Invitation." *FOCUS on Geography* (2019), http://www.focusongeography.org/publications/articles/cairns/index.html.

Olstad, Tyra. "Visitor Perception, Place Attachment, and Wilderness Management in the Adirondack High Peaks." In *Explorations in PLACE Attachment*, edited by Jeffrey S. Smith, 133–148. London: Routledge, 2018.

Olstad, Tyra. *Zen of the Plains: Experiencing Wild Western Places*. Denton TX: University of North Texas Press, 2014.

Raymond, Christopher M., Marketta Kyttä, and Richard Stedman. "Sense of Place, Fast and Slow: The Potential Contributions of Affordance Theory to Sense of Place." *Frontiers in Psychology* 8 (2017): 1–14.

Snyder, Gary. *The Real Work: Interviews and Talks 1964–1979*. New York: New Directions, 1980.

Soini, Katriina, Hanne Vaarala, and Eija Poutaa. "Residents' Sense of Place and Landscape Perceptions at the Rural–Urban Interface." *Landscape and Urban Planning* 104 (2012): 124–134.

Tempest Williams, Terry. *Red: Passion and Patience in the Desert*. New York: Vintage Books, 2002.

Tuan, Yi-Fu. *Space and Place: The Perspective of Experience*. Minneapolis: University of Minnesota Press, 1977.

We Are the Arctic. "About Us." Accessed May 1, 2020. https://www.wearethearctic.org/about-us.

Williams, Daniel R., Michael E. Patterson, Joseph W. Roggenbuck, and Alan E. Watson. "Beyond the Commodity Metaphor: Examining Emotional and Symbolic Attachment to Place." *Leisure Sciences* 14 (1992): 29–46.

Worrell, Richard and Michael C. Appleby. "Stewardship of Natural Resources: Definition, Ethical and Practical Aspects." *Journal of Agricultural and Environmental Ethics* 12 (2000): 263–277.

2 "I Was Under No Necessity of Seeking My Bread"

Robinson Crusoe and the Stewardship of Resources in Eighteenth-Century England[1]

Chris Mounsey

When the plot of a novel puts a man on a desert island for twenty-eight years, it might seem obvious that the author's motivation for writing was to explore human survival, adaptability, and stewardship of the environment in which he finds himself. However, since very early Defoe studies, critics have wanted to read more into it. William Lee's *Life of Defoe* (1866) reads *Robinson Crusoe* as a disguised autobiography and bemoans that "what he intended to be only veiled, time has rendered obscure."[2]

Had Lee known it, Karl Marx had already begun to lift what he thought to be the veil of obscurity from *Robinson Crusoe* in his *Grundrisse* (wr. 1857–61, pub. 1939), and the late entry into the world of this text perhaps accounts for the last eighty years of Crusoe criticism, which begins steeped in Marxian economics[3] before shifting to nineteenth-century Marxist analysis of Crusoe's labor, then to mid-twentieth-century accounts of Crusoe's individualist capitalism, and on to more recent post-colonial interpretations. Yet Max Novak points out that "Before embarking on his exposition of *Robinson Crusoe* and the labor theory of value, Karl Marx noted that even Ricardo had his economic parable *à la* Robinson."[4]

This could date the economic interest in Defoe's novel as early as David Ricardo's *The Principles of Political Economy and Taxation,*[5] published in 1817. To this end, Ufuk Karagöz gives a useful discussion about which early economists did, may have, and did not read *Robinson Crusoe*.[6] Yet this all gives the meaning of the text to a tradition of reading, and ignores the intentions of the author.

The oft-ignored third part of Crusoe's adventures, *Serious Reflections during the Life and Surprising Adventures of Robinson Crusoe: with his Vision of the Angelick World*, is even less of a rollicking adventure than the first two rather turgid volumes.[7] Yet it is introduced with a clue to how to read its predecessors:

> As the Design of every Thing is said to be first in the Intention, and last in the Execution; so I come now to acknowledge to my Reader, That the present Work is not merely the product of the two first

DOI: 10.4324/9781003219064-4

Volumes, but that the two first Volumes may rather he called the Product of this: The Fable is always made for the Moral, not the Moral for the Fable.[8]

My reading attempts to reverse the expansive contemplation of *Robinson Crusoe* that moves from individualist economics outwards to Empire, in order to emphasise the moral of the trilogy, which is the object of Crusoe's inward meditation on the sense of responsibility that underlies his stewardship of the island. But we must begin with an explanation of the paradox of the vast body of Crusoe scholarship before we move onto the promise offered by Defoe's moral, in which he prompts his readers to remember their human responsibility for stewarding the environment.

Why People Have Read *Robinson Crusoe* as an Economic Model

Ian Watt's *Rise of the Novel*, the foundation of most Marxist interpretations of Defoe in the past eighty years, associates *Robinson Crusoe* with "individualism" in the title of its third chapter.[9] The term, Watt reminds us, is anachronistically employed[10]; however, the concept, he argues, is discernible in the novel since the society which Crusoe reconstructs on the island

> obviously depends on a special type of economic and political organisation and on an appropriate ideology; more specifically, on an economic and political organisation which allows its members a very wide range of choices in their actions, and on an ideology primarily based, not on the tradition of the past, but on the autonomy of the individual, irrespective of his particular social status or personal capacity.[11]

Watt dates this economic and political organisation to the Glorious Revolution of 1689, remarking that, with other writers of the period, "Defoe somewhat ostentatiously set[s] the seal of literary approval on the heroes of economic individualism."[12]

It is not surprising that Ian Watt should add Daniel Defoe to his list of individualists who ostensibly pioneer the novel with a discussion of the exciting possibilities inherent in trade.[13] Defoe had been the editor of, and largely the writer of, the *Review of the State of the British Nation* for ten years (1703–1713), in which he waxed eloquently about 'trade'[14] over many numbers, explaining to his readers this new form of money making, which Karl Marx described as "the sphere of the circulation of commodities."

Ian Watt feels able to transfer his argument to *Robinson Crusoe* on a perceived shift from homemade necessities to a trade in the same goods as commodities, which resulted in "increased feminine leisure and the development of economic specialisation," a phenomenon he claims was observed by visitors from other European countries, Cesar de Saussure and Pehr Kalm: "The old household duties of spinning and weaving, making bread, beer, candles and soap, and many others, were no longer necessary, since most necessities were now manufactured and could be bought in shops and markets."[15]

This argument is so fundamental to *The Rise of the Novel* that it appears in the first chapter, "Realism and the Novel," and sets the foundation for Watt's argument for the popularity of what is, in the first volume, an extended account of Crusoe's household chores, presumably because there was no woman on the island to do them for him. Watt glosses this argument in his chapter on *Robinson Crusoe*, observing that

> When Crusoe makes bread, for instance, he reflects that "'Tis a little wonderful and what I believe few people have thought much upon, viz., the strange multitude of little things necessary in the providing, procuring, curing, dressing, making and finishing this one article of bread." Defoe's description goes on for seven pages, pages that would have been of little interest to people in medieval or Tudor society, who saw this and other basic economic processes going on daily in their own households.[16]

This is simply not true. Baking bread and other sweetmeats had been a major part of English business and trade from the early Middle Ages. So much so that the Worshipful Company of Bakers tells us on its website that: "The Bakers' Company can trace its origins back to 1155 and is the City of London's second oldest recorded guild."[17]

The same can be said of the other household necessities Watt lists – beer, candles, and soap – as the Worshipful Company of Brewers date their history from the late Middle Ages, suggesting that this production had already moved out of the sphere of domestic female responsibilities long before the eighteenth century: "The Company received its first charter from Henry VI in 1438 when the brewers were incorporated as 'The Wardens and Commonalty of the Mystery or Art of Brewers of the City of London.'"[18] As does the Worshipful Company of Wax Chandlers:

> In 1371 the Company gained Ordinances that gave them control over the trade of Wax Chandlers in the City of London. It is now governed by a Royal Charter, which was granted in 1484 – we are the only Livery Company to have a charter of King Richard III.[19]

The Worshipful Company of Tallow Chandlers, who made soap likewise dates from an era long before the one Watt associates with the "rise" of the novel: "The Worshipful Company of Tallow Chandlers has its origins around 700 years ago, with a group of craftsmen working together to support the tallow candle trade before being granted full Livery status in

1462."[20] Thus, there is no reason for us to believe Watt's assertion that "Defoe could therefore expect his readers to be interested in the very detailed descriptions of the economic life which comprise such an important and memorable part of his narrative."[21]

"Feminine leisure," if it ever existed, which was caused by a rise in the commercial production of the "old household [necessities such as] spinning and weaving,[22] making bread, beer, candles, and soap," is simply not an aspect of early eighteenth-century trade practices, conveniently allowing the novel to rise and fill in the gap in women's lives. There has to be another reason for the popularity of the novel.[23] Once we delink the rise of the novel from male economic individualism and ostensible female leisure, we can return it to the Christianity of Defoe's own era, as well as to Defoe's own deeply felt Presbyterian morality; in search of Defoe's moral, I shall turn to the subject of the third volume of the Crusoe trilogy.

Towards a Notion of Christian Stewardship

As do most other critical works on the Robinsonade, so shall this essay eschew the finer details of *Serious Reflections*, but rather center upon the first volume describing Crusoe's time on the island, while at the same time remembering that Defoe was writing from a religious perspective, such that the making of bread is the foremost aspect of the early part of the novel. The description of making bread is not a mere seven pages long, but more like ninety in the first edition, and it marks, I shall argue, not the political contract between citizens of John Locke's *Two Treatises of Government*[24] but the ideal relationship between humans and the environment in which they find themselves. Thus, this essay will read *Robinson Crusoe* as the story of a return to the roots of Christianity. Clues abound, as Genesis tells us, after eating of the fruit of the tree of the knowledge of good and evil: "Unto Adam and to his wife did the Lord God make coats of skins, and clothed them."[25]

If this may be accepted as a rather tongue-in-cheek source of Crusoe's famous clothes, we might also accept that *Robinson Crusoe* is an explanation of the paradox of Eden. Adam has eaten of the fruit of the knowledge of good and evil, and become god-like in wisdom, so has to be prevented from eating the fruit of the tree of life, by which he shall become immortal, and become equal with God: "And the Lord God said, Behold, the man is become as one of us, to know good and evil: and now, lest he put forth his hand, and take also of the tree of life, and eat, and live for ever."[26] This is why God banishes Adam and Eve from the Garden, and to do what? "Therefore the Lord God sent him forth from the garden of Eden, to till the ground."[27]

These verses act as an explanation as to why people, creations of an all-powerful God, are born and die, and must strive with the earth and its climate to grow food to eat. Had Adam and Eve remained in Eden, they would have been able to live on the fruits of the garden. It was their sin against the law of God:

> And the Lord God commanded the man, saying, Of every tree of the garden thou mayest freely eat: But of the tree of the knowledge of good and evil, thou shalt not eat of it: for in the day that thou eatest thereof thou shalt surely die.[28]

Thus sin, and the expulsion from Eden, set in motion the human life cycle, and their striving with the yearly cycle of planting and reaping, in the diurnal cycle of waking, working the land and sleeping. For this reason, I will argue that Defoe's trilogy was written to remind his readers that man does not live by commodities alone, and that bread is the staff of life (since beer, candles, and soap are three items which Crusoe fails to make on the island). But the emphasis on bread is probably because God tells Adam and Eve: "In the sweat of thy face shalt thou eat bread."[29] Only when the supply of bread has been secured can there be time for making commodities, and the crops to make bread can only be grown if balance is maintained between human intervention and the climate, and the fertility of land which is planted and sown. In this way, we might read *Robinson Crusoe* as an early example of the recognition that the need to take care of the environment is a prerequisite to more genteel living.

The history of my argument can be found in a slender thread running through *Robinson Crusoe* criticism. In a development of Ian Watt's economic argument, Roy J. Ruffin argues that Crusoe on his island offers the most basic form of economics even with a society of only one:

> Crusoe must choose between work and leisure; he must decide on how his working time is to be allocated between various goods; and he must solve an optimal capital accumulation programme. But we can do much more with a Crusoe economy: it is possible to simulate the workings of a price system.[30]

Following the simplest of economic models, where Crusoe is both producer and consumer, Ruffin explores "the nature of a competitive equilibrium and the classical theorems of welfare economics."[31] In so doing, Ruffin puts forward two models, one without pollution, and one with; he asks us to "imagine the production process emits a smelly substance that invades the entire island."[32] While the intention of Ruffin's essay is a thought experiment which has more to say about economic theory than stewardship, it does set in train the idea that human intervention into a largely pristine environment is not necessarily beneficial to the environment.

H. Daniel Peck takes a more realist view and confronts Watt's use of Crusoe's Island as a mythical structure on which to base theories of capitalism or the Protestant work ethic. Instead, Ruffin argues "the importance of environment on the development of the self," and maintains that "What is required... is an approach which maintains the environment as a real place, but at the same time deals with the problem of the self." Thus, Peck argues that the island is "a specific and positive landscape insofar as Crusoe enters into a willing relationship with it." The relationship Peck describes is one in which the use of tools is paramount in transforming Crusoe's opinion that "this environment is not totally resistant to his efforts, and that he can, in fact, create a tolerable way of life."[33] The relationship is therefore successful when it is one of human control and honing of the land as Crusoe hones his axe blades.

Emmanuelle Peraldo makes a similar argument, although she regards Crusoe's relationship with the island as more dynamic. Thus, the landscape in the first volume of Robinson Crusoe "is not merely a setting for his characters but fully integrated and continuous with them, in a single interconnected realm." Nevertheless, Peraldo argues "that landscape in *Robinson Crusoe* is a place on which Crusoe imprints himself to appropriate or master it."[34]

Steve Mentz addresses the difficulties of mastery of the environment, which lies at the heart of much post-colonialist criticism, that all human relationships with the earth are about conquering and enslaving, whether it be of the earth or of other races. Reading Crusoe's struggle from the wreck to the beach, Mentz argues, against realist readings, that "Crusoe's shipwreck points to a symbolic renovation of swimming as a way of responding to eco-catastrophe."[35] The image of Crusoe swimming within an environment which swirls around him in a dynamic relationship with him is really helpful. The world's environment is like a storm at sea, something within which we might find our way, by swimming in and through its currents and waves. There can be no mastery over a storm.

James Robert Wood goes some way to landing this dynamic relationship between human and environment, when he argues that

> The island on which Robinson Crusoe lives for twenty-seven years is strewn with things – caps, shoes, pots, coins, sheepskins, bones, and guns – and critics have had much to say about them. In this essay, however, I cast an eye down to the earth that lies beneath these things.[36]

Yet the essay does not look down for long, since for Wood, "it is by seeing Crusoe as an earthy creature in dynamic interaction with his earthy environment that the three books emerge most clearly as visions of empire, not by holding Crusoe superior to and separate from his surroundings."[37]

But, for Defoe, is Crusoe really superior to and separate from his surroundings? It is in the long description of making bread that we find that he is part of the island, albeit a new part.

The care Crusoe takes in making a loaf of bread, and the context in which he makes it, becomes a metaphor for the relationship between each human reader and the place in which they find themselves. Defoe thus reminds us to take the same care in our relationship with whatever place we find ourselves in the world and with whatever things we do to get our bread. Defoe's concerns about commodities, in other words, rely upon a continuing relationship between each human being and the earth on which the necessities of our lives are grown.

Making Bread

The famous way in which Crusoe begins the process of bread making bears detailed explanation. He has no more candles, so he makes a lamp with goat tallow and an oakum wick, during which task, "it happened that rummaging my Things, I found a little Bag, which I hinted before, had been filled with Corn to feed the poultry." Crusoe finds nothing useful in the bag, so shakes out its dusty contents "on one Side of my Fortification, under the Rock." The rains come, and Crusoe soon notices that the "few Stalks of something Green shooting out of the Ground" are "English Barley."[38] It is worth noting that all four necessities of life which Ian Watt believed were commercialised in the early eighteenth century are encompassed in this little scene: candles, baking and brewing (by the barley), and tallow for soap. But even more important about the barley is that Crusoe "threw the Stuff away."[39]

Crusoe's thoughtless act recalls two vital elements of Christianity: the rejected stone and the parable of the sower. Reminding his disciples that he is not going to be well received, Jesus uses an architectural metaphor: "Did ye never read in the scriptures, The stone which the builders rejected, the same is become the head of the corner: this is the Lord's doing, and it is marvellous in our eyes?"[40] Crusoe's throwing out of the "Husks and Dust" which eventually grow into his crop of barley is a similar act of rejection and one which leads him to both his spiritual saviour and to the "staff of life." But in the same way that Adam and Eve must learn how to farm after being cast out of Eden, so in order to reach his as yet unthought of goals, Crusoe must also learn how to steward his resources, as is made clear in the parable of the sower:

> As he was scattering the seed, some fell along the path, and the birds came and ate it up. Some fell on rocky places, where it did not have much soil. It sprang up quickly, because the soil was shallow. But when the sun came up, the plants were scorched, and they withered because they had no root. Other seed fell among thorns, which grew up and choked the plants. But other fell into good ground, and brought forth fruit, some an hundredfold, some sixtyfold, some thirtyfold. Who hath ears to hear, let him hear.[41]

Making sure that the grains are not haphazardly strewn is not a difficult lesson, but, as the continuation of the Parable of the Sower goes on to explain, stewardship must also be a conscious act of care:

> And the disciples came, and said unto him, Why speakest thou unto them in parables? He answered and said unto them, Because it is given unto you to know the *mysteries* of the kingdom of heaven, but to them it is not given. For whosoever hath, to him shall be given, and he shall have more abundance: but whosoever hath not, from him shall be taken away even that he hath.[42]

Crusoe's (and Defoe's readers') coming to consciousness of what he is (they are) doing, is, I would argue, why the growing of barley is interspersed with the earthquake, Crusoe's illness, and the discovery of the Edenic garden of grapes, lemons, and limes.

Crusoe begins as self-centered and with thoughts only of his ill-luck in being confined on the miserable island after the shipwreck, so for him to learn the mysteries of God's creation, that is, to understand his position on the island as steward of the land outside Eden is a long journey. At once, when Crusoe notices the growing stalks of barley, he thinks them a "Prodigy of Nature" and searches the island for more, believing they are "the pure Productions of Providence for my support."[43] This is the self-centered state of ignorance of his environment in which the castaway begins. He expects to find English barley on a Caribbean island, as though it has been put there for his use, as if God provided the makings of bread for him simply because he is an Englishman, as though every part of the world was a table spread with English comestibles simply because there is an Englishman to consume them.

Crusoe's lesson in stewardship begins when he remembers that emptying the "Bag of Chicken's Meat"[44] was the source of the alien corn, a memory which turns his thoughts presently away from God's Providence. Yet, looking back from the future, Crusoe reconstructs the real miracle:

> for it was really the work of Providence as to me, that should order or appoint that 10 or 12 grains of Corn should remain unspoil'd, (when the Rats had destroy'd all the rest), as if it had been dropt from Heaven; as also, that I should throw it out in that particular place, where, it being in the Shade of a high Rock, it sprang up immediately; whereas, if I had thrown it anywhere else at that Time, it had been burnt up and destroy'd.[45]

Second sight of events is another small miracle of Providence brought to us in this realist narrative: Defoe uses the fast-disappearing ink as a method of allowing a clash of reflections between the journal he writes on the island with the small resources of ink and the narrative he writes when he returns to England: present and future. And the clashes pile up at this turning point in the text, Pelion upon Ossa, the most important being the projection of the first loaf of bread which is portended by the barley:

> I carefully saved the ears of this Corn, you may be sure, in their Season, which was about the End of June; and, laying up every Corn, I resolved to sow them all again, hoping in Time to have some quantity sufficient to supply me with Bread. But it was not till the 4th Year that I could allow myself the least Grain of this Corn to eat, and even then but sparingly, as I shall say afterwards, in its order; for I lost all that I sowed the first Season by not observing the proper time; for I sowed it just before the dry Season, so that it never came up at all, at least not as it would have done; of which in its Place.[46]

Acting as parentheses around Crusoe's journey to stewardship is his making bread. In the passage above we find that he must carefully calculate the seasons of the Caribbean if he is to plant his grain at an opportune moment, but there are other lessons which are required simultaneously to bring him to conscious stewardship of his island. First is the knowledge that the wall he has built to ensure his safety from attack by marauders or wild beasts on the island offers him no security, when an earthquake strikes "as would have overturned the strongest Building as could be suppos'd to have stood on this Earth."[47] This is his first reminder that the environment is more powerful than he, and that he must work within its scope since it will not submit to his will: the barley will not grow out of season, the wall will not withstand the earthquake.

Nor is the wall he has constructed all beneficial in the protections it offers. Contemporary science of the earthquakes connected shaking earth with climatic conditions and when the expected rains come Crusoe is forced "to cut a hole through [his] new fortification, like a sink, to let the water go out, which would else have flooded my cave."[48] His dilemma leads him to desire to remove to another part of the island, but with the realisation that it will take "a vast deal of time" to do this, he resolves to build a camp elsewhere of "Piles and Cables,"[49] that is posts and ropes, before he quits his perilous safe house.

The work needed to build the new dwelling requires tools, and tools require honing, which takes "two whole Days,"[50] during which time, Crusoe notices that the biscuits he has brought from the wreck are running out. Crusoe calls his hard tack "Bread," which suggests the importance of learning to grow his barley successfully, so he can replenish his supply of this necessity of life. Yet at the same time, the wreck has moved so he spends every day but one from May 3 to June 15 cutting planks of wood from it and bringing ashore lead and iron bolts. His salvage operation leaves him little time to get food, the day he spends fishing is a failure, and except for the fortuitous arrival of a misdirected turtle to his beach, which he consumes along with its "threescore eggs,"[51] he is still surviving on the diminishing stock of hard tack.

Crusoe's food resources continue to decline when, either due to eating the turtle and its eggs, the heavy rain, or some local infection he becomes ill for ten days, whence he sees a vision of a flaming man descending in a cloud of fire who tells him "Seeing all these Things have not brought thee to Repentance, now thou shalt die."[52] For the first time in his life, Crusoe begins to reflect on his past actions and wondering whether his isolation was due to "the Hand of God against me."[53]

His ruminations on his uncertain future remind him what the fortuitous growing of the barley had not taught him:

> The growing up of the Corn, as is hinted in my Journal, had at first some little Influence upon me, and began to affect me with Seriousness, as long as I thought it had something miraculous in it; but as soon as ever that Part of the Thought was removed, all the impression that was raised from it wore off also, as I have noted already.[54]

He then notes what the earthquake had not taught him:

> Even the Earthquake, though nothing could be more terrible in its Nature, or more immediately directing to the invisible Power which alone directs such Things, yet no sooner was the first Fright over, but the Impression it had made went off also. I had no more sense of God or His judgments – much less of the present Affliction of my Circumstances being from His hand – than if I had been in the most prosperous Condition of Life.[55]

Only thoughts of his imminent death can awaken his conscience to feel he has "provok'd the Justice of God."[56] But we must not forget that this religious awakening has been circumscribed by the growing of food and the unruliness of the environment, and it is to these aspects of his life that the newly religious Crusoe turns his efforts, and away from the wreck. He must not look back to his former life, but forward to his life on the island, to provide him with continued support.

Yet the journey to a constant supply of food is still not straight. When he wakes from the fever dream, Crusoe measures his weakness in being unable to carry his gun, the weapon being an old sign of his fear of attack, though he is alone on an island that is home to no fiercer animal than a wildcat. This represents his old self, but his ruminations suggest the new:

> As I sat here some such Thoughts as these occurred to me: What is this Earth and Sea, of which I have seen so much? Whence is it produced? And what am I, and all the other Creatures Wild and Tame, Humane and Brutal? Whence are we? Sure we are all made by some secret Power, who formed the earth and sea, the air and sky. And who is that?

Then it followed most naturally, it is God that has made all. Well, but then it came on Strangely, if God has made all these things, He guides and governs them all, and all things that concern them; for the Power that could make all things must certainly have Power to guide and direct them. If so, nothing can happen in the great Circuit of His works, either without His Knowledge or Appointment.[57]

I draw attention to this brief explanation of God and His creation as it tells us that Crusoe's old self-centeredness has not gone away completely. The feeling that Crusoe has a personal God, who has turned His eye specifically to him, allows him to think himself out of the situation in which he finds himself, since he has not yet remembered that humans have been excluded from paradise, and must sweat for their bread. Thus, on reading the Bible for the first time, Crusoe, pulls out quotations that put him back in the center of his own universe:

> *Call on me in the Day of Trouble, and I will deliver, and you shalt glorify me.*
> *Can God spread a Table in the Wilderness?*[58]

The first verse helps Crusoe to understand that the word "deliver" does not imply his imminent rescue from the island. The second, however, suggests that the island itself might be a personal Garden of Eden. When he finally explores his island "prison,"[59] Crusoe discovers the glories of its gardens, furnished with tobacco, but not cassava (so there is no means to make what he expects to be the local bread), aloes (which he does not at first understand), grapes, cocoa trees, lemons, and limes. Thus, he surveys the vale

> with a secret kind of pleasure,... [and he begins] to think that this was all my own; that I was King and Lord of all this Country indefeasibly, and had a Right of possession; and if I could convey it, I might have it in Inheritance as completely as any Lord of a Manor in *England*.[60]

This passage has oft been quoted to confirm Crusoe's colonization of the island and sets up many of the post-colonialist readings of the novel.[61] I would argue rather that his claim of kingship marks Crusoe's old selfish self still speaking, still with his belief that his personal God will help him, an Adam in a new Eden. He must still learn that his forefather has been excluded from Eden and that he must sweat for his bread. He cannot live by picking the fruit of the trees as Adam did, even though Crusoe "was warned by [his] Experience to eat sparingly of [the grapes]; remembering that when [he] was Ashore in Barbary, the eating of Grapes killed several of our Englishmen, who were Slaves there, by throwing them into Fluxes and Fevers."[62]

Likewise, when he tries to store up these un-sweated-for riches, by taking them to his "Home," he fails, and must work even to eat the freely given fruit:

> Accordingly, having spent three Days in this Journey, I came Home (so I must now call my Tent and my Cave); but before I got thither the Grapes were spoiled; the richness of the fruit and the weight of the Juice having broken them and bruised them, they were good for little or nothing; as to the Limes, they were good, but I could bring but a few. The next Day, being the 19th, I went back, having made me two small Bags to bring home my Harvest; but I was surpriz'd, when coming to my Heap of Grapes, which were so rich and fine when I gather'd them, to find them all spread about, trod to pieces, and drag'd about, some here, some there, and Abundance eaten and devour'd. By this I concluded there were some wild creatures thereabouts, which had done this; but what they were I knew not.
>
> However, as I found there was no laying them up on Heaps, and no carrying them away in a Sack, but that one way they would be destroy'd, and the other way they would be crushed with their own Weight, I took another course; for I gathered a large quantity of the grapes, and hung upon the out-branches of the Trees, that they might cure and dry in the sun; and as for the Limes and Lemons, I carried as many back as I could well stand under.[63]

In making raisins of the grapes, Crusoe changes almost imperceptibly from being King in Eden to learning stewardship of the gifts of the island. In this way, he intuits that the easily won pleasures of the fruits of the "tempting" garden will not be good for him, and he thus decides not to "remove [his] Habitation" thence lest he succumb to the lure of kingship with its dainty but potentially lethal offerings. Instead, he builds himself "a kind of a Bower," where he stays only "two or three nights together."[64] It is no surprise, then, that simultaneously, Crusoe decides to choose a day for the Sabbath, which he marks on his diary stick with a "longer Notch."[65] It is of no surprise either that this is the time that his ink begins to run out, and nor is it surprising that this is the moment that he begins to grow barley with the intention of making bread. Crusoe's stewardship is all of a piece: he must take care of himself in the present and keep an eye on his future, and the process of making bread brings all these together.

On first sowing, Crusoe is uncertain of the "proper Time"[66] as he has not worked out the wet and dry seasons of the island: something which his notched stick cannot calculate. He must learn the yearly cycle of the climate from the equinoxes, and plant seed just before the rains come.

Using the passive form of the verb, Crusoe tells his readers, "by this Experiment I was made Master of my Business, and knew exactly when the proper Season was to sow, and that I might expect two Seed-times and two Harvests every year."⁶⁷ Albeit the word "Master" seems to suggest something like political control, yet the verb in its passive form – "I was made Master" – more likely refers to the idea that Crusoe has served an apprenticeship and has learned how to be master of the mystery of farming. The idea of mastering the "mystery" of a trade is the language of the London Companies who regulated apprenticeships.

Crusoe's mastery of farming is not completed with his knowledge of the seasons, and he is yet to learn that in order to be a good steward he must build fences to keep out the hares, as well as scare the birds away if he wants to increase his yield. Likewise, he has to master the mystery of basket weaving to store the barley grains as his stock becomes large enough to allow him to use some to make bread.

This new stewarding personality begins to take over all aspects of his life, and ensures that he does not shoot as many birds as he can as he has to be "sparing of [his] Powder."⁶⁸ Likewise, the change encourages him – after a year of fear of being is what he considered to be a miserable place– to reconnoitre and map the whole island to discover its secrets.

It is important that Crusoe describes the change within himself as a function of place. It is not at all that he becomes master of the island. Mastery, as suggested above, is the language of gaining skills. Thus, he rejects the idea of removing his habitation to the coastal region which supports a large colony of birds and turtles, which would provide food with little work and require a lot of gunpowder that he does not have. Rather, as Crusoe learns to work with his environment, so he becomes at one with himself. This is not the same as his yearning for or recreating his home in England (or Canada), discussed by a number of critics, but a realisation that if he can master the skills required to fulfil the requirements of his life on this Caribbean island, then he might as well live anywhere.⁶⁹ He describes his change thus:

> Before, as I walked about, either on my hunting or for viewing the country, the anguish of my soul at my condition would break out upon me on a sudden, and my very heart would die within me,...
> From this Moment I began to conclude in my Mind that it was possible for me to be more happy in this forsaken, solitary Condition than it was probable I should ever have been in any other particular State in the World.⁷⁰

The island is still a solitary place, but his wide skill set, including carpentry, bread making, hunting, basket weaving and farming, enables him to feel a harmony with his environment. He can turn wood into shelves, barley into bread, osiers into baskets, and the fowls of the air and the turtles of the sea into dinner.

As of yet he cannot make earthenware pots, soap, clay pipes, beer, or glass, but the illumination that the skills he has not yet mastered require no more than appropriating mysteries that he may yet encompass, brings him into an even more comfortable relationship with creation. The process of learning is far from re-creating a new England, or one of making or desiring commodities – when he does make a successful earthenware pot it is ugly, but it works to contain – but more one of stewarding the resources at hand with thoughts of maintaining his future in that place and in a like manner. Happiness in the present requires a sense that the future is taken care of.

In his third year on the island, Crusoe now can sow his barley, protect it from other contenders for eating it, reap it (with a sword – there has to be a joke here about not being able to beat his sword into a ploughshare, let alone a sickle). Yet as Crusoe puts it, "It can be truly said, that now I work'd for my Bread,"[71] which fulfils the requirement of Genesis. In his fourth year, Crusoe makes a pestle and mortar, and a sieve to make his flour, so that he reaches another goal when he masters the mysteries of baking, when "in the best Oven in the world, I baked my barley-loaves, and became in little time a good Pastry-Cook into the bargain; for I made myself several Cakes and Puddings of the Rice."[72]

Perhaps the most important aspect of Crusoe's growing barley is that it never becomes a capitalist business, producing excess. Crusoe writes:

> And now, indeed, my Stock of Corn increasing, I really wanted to build my Barns bigger; I wanted a place to lay it up in, for the Increase of the Corn now yielded me so much, that I had of the Barley about twenty Bushels, and of the rice as much or more; insomuch that now I resolved to begin to use it freely; for my Bread had been quite gone a great While; also I resolved to see what Quantity would be sufficient for me a whole Year, and to sow but once a Year.
> Upon the whole, I found that the forty Bushels of Barley and Rice were much more than I could consume in a Year; so I resolved to sow just the same Quantity every year that I sowed the last, in hopes that such a Quantity would fully provide me with Bread, &c.[73]

To plant just enough for survival without producing an excess, challenges the idea that Crusoe's economy is capitalist, and furthermore, the idea of gaining a wide skill set challenges the idea that Crusoe is setting up a commodity-based market. He has no-one to sell his wares to: there is no "trade" or market.

The Rise of the Economic Model

The foregoing has been an attempt to suggest a corrective to the Marxist and colonialist inspired readings of *Robinson Crusoe* that have been predominant for the past eighty years. However, there is little doubt that capitalism did develop before and during the eighteenth century with the onset of trade between various nations of the world, and the setting up of such trading groups as the East India Company (1600–1864). This company was unlike the other London Companies with which the chapter began, since its business was not to impart skills and impose discipline upon its members for failing to meet the required standards. Rather, the East India Company was a "closed shop" which demanded membership as a license to trade with any but European countries.[74] It began with little more than a group of pirates who stole the exotic cargoes of other European ships; as the Company developed, however, it began to set up factories in various places in the Pacific rim and India, whence it did its own trade in commodities such as spices, porcelain, and other consumer desirables that were hard to source in Britain. I would argue that it is on this development, that the Marxist and colonialist understanding of early eighteenth-century trading culture should more properly have been based, particularly since it must be remembered that Daniel Defoe was no successful capitalist. He had gone bankrupt twice, in 1696 and 1706, due to his poor trading skills, and lived the rest of his life under the burden of £17,000 of debt because of it.[75] This continuing state of affairs, under which Defoe wrote his novel, seems to me to be a much more important guide to *Robinson Crusoe* (if I may be allowed to lift the veil of the novel and find something more following William Lee's expectations of it). It would suggest the reason for Crusoe not planting more barley than was necessary for his own consumption: capitalism was a precarious venture which could land investors in serious financial trouble.

This is not to argue that there were no developments in farming techniques which ran concomitantly with the development of world trade, to which Defoe or Karl Marx might have been referring in their work. I have argued elsewhere that farming practice in the late seventeenth century was politically motivated, largely concerning the question of whether to use "hot manure," which Virgil suggested in his Georgics, or well-rotted manure, which was the new practice of the Commonwealth, expounded on by John Worlidge.[76] Robinson Crusoe does not tell us which type of manure he uses, but the fact that he does manure his crop[77] suggests that Defoe was up to date with late seventeenth-century thinking about soil care.

It was not until at least ten years after the publication of *Robinson Crusoe* that capitalism can be associated with farming in what is now called the Agricultural Revolution. Charles "Turnip" Townshend, second Viscount Townshend did not begin work on the Norfolk four course system until after he retired from government on May 15, 1730. Nor was Jethro Tull's system well known before the publication of his *Horse Hoeing Husbandry* in 1731.[78]

It might be argued that Defoe's *A Tour thro' the Whole Island of Great Britain*, which dates from 1724, suggests that the author of our novel might have known of the coming revolutionary tactics that would soon be imposed upon agriculture, albeit his tour of the larger island was published five years after the description of Robinson Crusoe's. Yet I can locate nothing more than references to the quality of the wheat and barley grown in various British regions, and nothing on farming techniques.

For this reason, I would suggest that Daniel Defoe's novel is more of a strong recommendation to the stewardship of the small island of Britain, than a spur to world trade and domination. Robinson Crusoe sweats for the bread he eats and learns the appropriate skills to provide other items to make his life a little easier. This puts him on an equal footing with the environment in which he lives and moves and has his being. Crusoe becomes a part of the island's environment: "And what is this Earth and Sea, of which I have seen so much?... And what am I, and all the other Creatures Wild and Tame, Humane and Brutal?"[79] To which the island has answered: you have become part of the island, and have learned to cherish me.

Notes

1 Daniel Defoe, *The Life and Strange Surprizing Adventures of Robinson Crusoe, of York Mariner* (London: W. Taylor, 1719), 4.
2 William Lee, *Daniel Defoe: His Life and Recently Discovered Writings*, 3 vols., vol 1 (London: John Camden Hotten, 1869), 2.
3 For example: Maximillian E. Novak, "Robinson Crusoe's Fear and the Search for Natural Man," *Modern Philology* 58, no. 4 (May 1961): 238–245; Maximillian E. Novak, "Robinson Crusoe's 'Original Sin,'" *Studies in English Literature, 1500-1900* 1, no. 3 *Restoration and Eighteenth Century* (Summer, 1961): 19–29; Gustav Hübener, "Das Kaufman Robinson Crusoe," *Englische Studien*, LIV (1920): 367–398; Brian Fitzgerald, *Daniel Defoe* (London, 1954), 60–92; Ian Watt, *The Rise of the Novel* (London, 1957), 60–92; Maximillian E. Novak, "Crusoe the King and the Political Evolution of His Island," *Studies in English Literature, 1500-1900* 2, no. 3 (Summer 1962): 337–350; Frank Donoghue, "Inevitable Politics: Rulership and Identity in *Robinson Crusoe*," *Studies in the Novel* 27, no. 1 (Spring 1995): 1–11; Lydia H. Liu, "Robinson Crusoe's Earthenware Pot," *Critical Inquiry* 25, no. 4 (Summer, 1999): 728–757; Wolfram Schmidgen, "Robinson Crusoe, Enumeration, and the Mercantile Fetish," *Eighteenth-Century Studies* 35, no. 1 (Fall, 2001): 19–39; Ann Van Sant, "Crusoe's Hands," *Eighteenth-Century Life* 32, no. 2 (Spring 2008): 120–137; Lynn Festa, "Crusoe's Island of Misfit Things," *The Eighteenth Century* 52, no. 3/4, The Drift of Fiction: Reconsidering the Eighteenth-Century Novel (Fall/Winter): 443–471; David Wallace Spielman, "The Value of Money in *Robinson Crusoe, Moll Flanders,* and *Roxana*," *The Modern Language Review* 107, no. 1 (January 2012): 65–87; Jason H. Pearl, "Desert Islands And Urban Solitudes In The 'Crusoe' Trilogy," *Studies in the* Novel 44, no. 2 (Summer 2012): 125–143.
4 Maximillian E. Novak, "Robinson Crusoe and Economic Utopia," *The Kenyon Review* 25, no. 3 (Summer 1963): 474, 474–490.
5 David Ricardo, *The Principles of Political Economy and Taxation* (London: John Murray, 1817).

6 Ufuk Karagöz, "The Neoclassical Robinson: Antecedents and Implications," *History of Economic Ideas* 22, no. 2 (2014): 75–100.
7 Daniel Defoe, *Serious Reflections during the Life and Surprising Adventures of Robinson Crusoe: with his Vision of the Angelick World* (London: W. Taylor, 1720).
8 Defoe, *Serious Reflections*, A2.
9 Ian Watt, *Rise of the Novel* (Berkeley: University of California Press, 1957), kindle edition. As a blind scholar I am constrained to use electronic texts, often lacking page numbers; few page numbers are given, therefore, as they are meaningless to me. To find a reference, sighted scholars may use the electronic text that they find amenable, and search for the words using that function.
10 Watt writes: "It is very difficult to say when this change of orientation began to affect society as a whole – probably not until the nineteenth century. But the move certainly began much earlier. In the sixteenth century the Reformation and the rise of national states decisively challenged the substantial social homogeneity of mediaeval Christendom, and, in the famous words of Maitland, 'for the first time, the Absolute State faced the Absolute Individual.'" *Rise of the Novel*.
11 Watt, *Rise of the Novel*.
12 Watt, *Rise of the Novel*.
13 Since Watt's *Rise of the Novel* appeared in 1957, many other scholars have offered corrections to his "triple rise" theory of individualist capitalism, individualist Whig political philosophy, and individualist Protestant theology aligning with the individualist focus of the novel. Michael McKeon in 1987 offered a more dialectical approach to the emergence of progressive ideology from the aristocratic (Michael McKeon, *Origins of the English Novel* (Baltimore: Johns Hopkins, 1987)). Jane Spencer added the history of women novelists to the mix (Jane Spencer, *The Rise of the Woman Novelist* (Oxford: Blackwell, 1986)), and Rachel Carnell traced a less individualistic history of Whig, Tory, and Jacobite novels, in contrast to Watt's focus on the Whig individual as the central story of novelistic history (Rachel Carnell, *Partisan Politics, Narrative Realism and the Rise of the Novel* (New York: Palgrave Macmillan, 2006)).
14 In the *Review*, Defoe uses the word "trade" more than 6000 times.
15 Watt, *Rise of the Novel*.
16 Watt, *Rise of the Novel*.
17 https://www.bakers.co.uk.
18 https://www.brewershall.co.uk/.
19 https://www.waxchandlers.org.uk/about/.
20 https://www.tallowchandlers.org/.
21 Watt, *Rise of the Novel*.
22 The Drapers Company was granted its charter in 1364.
23 And, of course, it must be remembered that Ian Watt did not count women writing novels as significant, so he was exploring the rise of the novel in a history that neglected the first forty or so years of its progress, overlooking the novels of Aphra Behn and Eliza Haywood, among others.
24 Suggested by Watt, *Rise of the Novel*.
25 Genesis 3: 21.
26 Genesis 3: 22.
27 Genesis 3: 23.
28 Genesis 2: 16–17.
29 Genesis 3: 19.

30 Roy J. Ruffin, "Pollution in a Crusoe Economy," *The Canadian Journal of Economics / Revue canadienne d'Economique* 5, no. 1 (February 1972): 110.
31 Ruffin, "Pollution in a Crusoe Economy," 110.
32 Ruffin, "Pollution in a Crusoe Economy," 112.
33 H. Daniel Peck, "*Robinson Crusoe*: The Moral Geography of Limitation," *The Journal of Narrative Technique* 3, no. 1 (January 1973): 20–23.
34 Emanuelle Peraldo, "'Two broad shining eyes': Optic Impressions and Landscape in Robinson Crusoe," *Digital Defoe* 4 (Fall 2012): 18, https://english.illinoisstate.edu/digitaldefoe/features/peraldo.pdf.
35 Steve Mentz, "'Making the green one red': Dynamic Ecologies in Macbeth, Edward Barlow's Journal, and Robinson Crusoe," *Journal for Early Modern Cultural Studies* 13, no. 3, Commons and Collectivities: Renaissance Political Ecologies (Summer 2013): 76.
36 James Robert Wood, "Robinson Crusoe and the Earthy Ground", *Eighteenth-Century Fiction* 32, no. 3 (Spring 2020): 381.
37 Wood, "Robinson Crusoe and the Earthy Ground," 383.
38 Defoe, *Robinson Crusoe*, 90.
39 Defoe, *Robinson Crusoe*.
40 Matthew 21:42. Also at Mark 12:10; Luke 20:17; Acts 4:11; Ephesians 2:20; 1 Peter 2:6.
41 Matthew 13:3–9. Also at Mark 4:3; Luke 8:5.
42 Matthew 13:10–13.
43 Defoe, *Robinson Crusoe*, 91.
44 Defoe, *Robinson Crusoe*.
45 Defoe, *Robinson Crusoe*, 92.
46 Defoe, *Robinson Crusoe*.
47 Defoe, *Robinson Crusoe*, 93.
48 Defoe, *Robinson Crusoe*, 95.
49 Defoe, *Robinson Crusoe*, 96.
50 Defoe, *Robinson Crusoe*, 97.
51 Defoe, *Robinson Crusoe*, 100.
52 Defoe, *Robinson Crusoe*, 102.
53 Defoe, *Robinson Crusoe*, 105.
54 Defoe, *Robinson Crusoe*.
55 Defoe, *Robinson Crusoe*.
56 Defoe, *Robinson Crusoe*.
57 Defoe, *Robinson Crusoe*, 107.
58 Defoe, *Robinson Crusoe*, 110.
59 Defoe, *Robinson Crusoe*, 113.
60 Defoe, *Robinson Crusoe*, 117.
61 Brett C. McInelly, "Expanding Empires, Expanding Selves: Colonialism, the Novel, and *Robinson Crusoe*," *Studies in the Novel* 35, no. 1 (Spring 2003): 1–21; Joshua Grasso, " An Enemy of his Country's Prosperity and Safety: Mapping the English Traveler in Defoe's *Robinson Crusoe*," *CEA Critic* 70, no. 2 (Winter 2008): 15–30; John C. Traver, "Defoe, Unigenitus, and the 'Catholic' Crusoe," *Studies in English Literature, 1500–1900* 51, no. 3 Restoration and Eighteenth Century (Summer 2011): 545–563; Peter Walmsley, "Robinson Crusoe's Canoes," *Eighteenth-Century Life* 43, no. 1 (January 2019): 1–23; Robert Markley, "Introduction: Leaving Crusoe's Island," *Eighteenth-Century Fiction* 32, no. 1 (Fall 2019): 1–8; Barbara Fuchs, "Crusoe's Absence," *Studies in Eighteenth-Century Culture* 49 (2020): 27–42; Sten Pultz Moslund, "Postcolonialism, the Anthropocene, and New Nonhuman Theory: A Postanthropocentric Reading of Robinson Crusoe," *ariel: A Review of International English Literature* 52, no. 2 (April 2021): 1–38.

62 Defoe, *Robinson Crusoe*, 116.
63 Defoe, *Robinson Crusoe*, 118
64 Defoe, *Robinson Crusoe*, 118–19.
65 Defoe, *Robinson Crusoe*, 122.
66 Defoe, *Robinson Crusoe*, 123.
67 Defoe, *Robinson Crusoe*.
68 Defoe, *Robinson Crusoe*, 129.
69 T. D. MacLulich, "Crusoe in the Backwoods: A Canadian Fable?" *Mosaic: An Interdisciplinary Critical Journal* 9, no. 2 *Literature And Ideas* (Winter 1976): 115–126; Hugh Jenkins, "Crusoe's Country House(s)," *The Eighteenth Century* 38, no. 2 (Summer 1997): 118–133; Robert P. Marzec, "Enclosures, Colonization, and the Robinson Crusoe Syndrome: A Genealogy of Land in a Global Context," *boundary* 29, no. 2 (Summer 2002): 129–156; Everett Zimmerman, "Robinson Crusoe and No Man's Land," *The Journal of English and Germanic Philology* 102, no. 4 (October 2003): 506–529; Kit Fan, "Imagined Places: Robinson Crusoe And Elizabeth Bishop," *Biography* 28, no. 1 Inhabiting Multiple Worlds: Auto/Biography In An (Anti-) Global Age (Winter 2005): 43–53; Susan Fraiman, "Shelter Writing: Desperate Housekeeping from 'Crusoe' to 'Queer Eye'," *New Literary History* 37, no. 2 Critical Inquiries (Spring 2006): 341–359.
70 Defoe, *Robinson Crusoe*, 132–33.
71 Defoe, *Robinson Crusoe*, 138.
72 Defoe, *Robinson Crusoe*, 145.
73 Defoe, *Robinson Crusoe*, 146.
74 Robert Kerr, *A General History and Collection of Voyages and Travels*.(London: W. Blackwood, 1813), 102. Anthony Farrington, *Trading Places: The East India Company and Asia 1600–1834* (London: British Library, 2002).
75 W. R. Owens, P. N. Furbank, "Defoe and Imprisonment for Debt: Some Attributions Reviewed," *The Review of English Studies*, New Series 37, no. 148 (November 1986): 495–502.
76 Chris Mounsey, *Christopher Smart: Clown of God* (Lewisburg: Bucknell University Press, 2001), chapter 4.
77 Defoe, *Robinson Crusoe*, 136.
78 Jethro Tull, *Horse Hoeing Husbandry* (London: G. Strahan in Cornhill; T. Woodward in Fleet-Street; A. Miller over-against St. Clement's-Church in the Strand; J. Stagg in Westminster-Hall; and J. Brindley in New-Bond-Street, 1731).
79 Defoe, *Robinson Crusoe*, 107.

Bibliography

Carnell, Rachel. *Partisan Politics, Narrative Realism and the Rise of the Novel.* New York: Palgrave Macmillan, 2006.

Defoe, Daniel. *Serious Reflections during the Life and Surprising Adventures of Robinson Crusoe: with his Vision of the Angelick World.* London: W. Taylor, 1720.

Defoe, Daniel. *The Life and Strange Surprizing Adventures of Robinson Crusoe, of York Mariner.* London: W. Taylor, 1719.

Donoghue, Frank. "Inevitable Politics: Rulership and Identity in *Robinson Crusoe*." *Studies in the Novel* 27, no. 1 (Spring 1995): 1–11.

Fan, Kit. "Imagined Places: Robinson Crusoe And Elizabeth Bishop." *Biography* 28, no. 1 Inhabiting Multiple Worlds: Auto/Biography In An (Anti-) Global Age (Winter 2005): 43–53.

Farrington, Anthony. *Trading Places: The East India Company and Asia 1600–1834*. London: British Library, 2002.

Festa, Lynn. "Crusoe's Island of Misfit Things." *The Eighteenth Century* 52, no. 3/4, The Drift of Fiction: Reconsidering the Eighteenth-Century Novel (Fall/Winter): 443–471.

Fitzgerald, Brian *Daniel Defoe* (London, 1954).

Fraiman, Susan. "Shelter Writing: Desperate Housekeeping from 'Crusoe' to 'Queer Eye'." *New Literary History* 37, no. 2 Critical Inquiries (Spring, 2006): 341–359.

Fuchs, Barbara. "Crusoe's Absence." *Studies in Eighteenth-Century Culture* 49 (2020): 27–42.

Grasso, Joshua. "'An Enemy of his Country's Prosperity and Safety': Mapping the English Traveler in Defoe's *Robinson Crusoe*." *CEA Critic* 70, no. 2 (Winter 2008): 15–30.

Hübener, Gustav. "Das Kaufman Robinson Crusoe." *Englische Studien* LIV (1920): 367–398.

Jenkins, Hugh. "Crusoe's Country House(s)." *The Eighteenth Century* 38, no. 2 (Summer 1997): 118–133.

Karagöz, Ufuk. "The Neoclassical Robinson: Antecedents and Implications." *History of Economic Ideas* 22, no. 2 (2014): 75–100.

Kerr, Robert. *A General History and Collection of Voyages and Travels*. London: W. Blackwood, 1813.

Lee, William. *Daniel Defoe: His Life and Recently Discovered writings*, 3 vols., vol 1. London: John Camden Hotten, 1869.

Liu, Lydia H. "Robinson Crusoe's Earthenware Pot." *Critical Inquiry* 25, no. 4 (Summer, 1999): 728–757.

MacLulich, T. D. "Crusoe in the Backwoods: A Canadian Fable?" *Mosaic: An Interdisciplinary Critical Journal* 9, no. 2, Literature And Ideas (Winter 1976): 115–126.

Markley, Robert. "Introduction: Leaving Crusoe's Island." *Eighteenth-Century Fiction* 32, no. 1 (Fall 2019): 1–8.

Marzec, Robert P. "Enclosures, Colonization, and the Robinson Crusoe Syndrome: A Genealogy of Land in a Global Context." *Boundary* 29, no. 2 (Summer 2002): 129–156.

McInelly, Brett C. "Expanding Empires, Expanding Selves: Colonialism, the Novel, and *Robinson Crusoe*." *Studies in the Novel* 35, no. 1 (Spring 2003): 1–21.

McKeon, Michael. *Origins of the English Novel*. Baltimore: Johns Hopkins, 1987.

Mentz, Steve. "Making the green one red: Dynamic Ecologies in Macbeth, Edward Barlow's Journal, and Robinson Crusoe." *Journal for Early Modern Cultural Studies* 13, no. 3, Commons and Collectivities: Renaissance Political Ecologies (Summer 2013): 66–83.

Moslund, Sten Pultz. "Postcolonialism, the Anthropocene, and New Nonhuman Theory: A Postanthropocentric Reading of Robinson Crusoe." *ariel: A Review of International English Literature* 52, no. 2 (April 2021): 1–38.

Mounsey, Chris. *Christopher Smart: Clown of God*. Lewisburg: Bucknell University Press, 2001.

Novak, Maximilian E. "Crusoe the King and the Political Evolution of His Island." *Studies in English Literature, 1500–1900* 2, no. 3 (Summer 1962): 337–350.

Novak, Maximillian E. "Robinson Crusoe and Economic Utopia." *The Kenyon Review* 25, no. 3 (Summer 1963): 474–490.

Novak, Maximillian E. "Robinson Crusoe's 'Original Sin'." *Studies in English Literature, 1500–1900* 1, no. 3, Restoration and Eighteenth Century (Summer 1961): 19–29.

Novak, Maximillian E. "Robinson Crusoe's Fear and the Search for Natural Man." *Modern Philology* 58, no. 4 (May 1961): 238–245.

Owens, W. R., and P. N. Furbank, "Defoe and Imprisonment for Debt: Some Attributions Reviewed." *The Review of English Studies*, New Series 37, no. 148 (November 1986): 495–502.

Pearl, Jason H. "Desert Islands and Urban Solitudes in the 'Crusoe' Trilogy." *Studies in the Novel* 44, no. 2 (Summer 2012): 125–143.

Peck, H. Daniel. "*Robinson Crusoe*: The Moral Geography of Limitation." *The Journal of Narrative Technique* 3, no. 1 (January, 1973): 20–23.

Peraldo, Emanuelle. "'Two broad shining eyes': Optic Impressions and Landscape in Robinson Crusoe." *Digital Defoe* 4 (Fall 2012): 18–30.

Ricardo David. *The Principles of Political Economy and Taxation*. London: John Murray, 1817.

Ruffin, Roy J. "Pollution in a Crusoe Economy." *The Canadian Journal of Economics / Revue canadienne d'Economique* 5, no. 1 (February 1972): 110–127.

Schmidgen, Wolfram. "Robinson Crusoe, Enumeration, and the Mercantile Fetish." *Eighteenth-Century Studies* 35, no. 1 (Fall 2001): 19–39.

Spencer, Jane. *The Rise of the Woman Novelist*. Oxford: Blackwell, 1986.

Spielman, David Wallace. "The Value of Money in *Robinson Crusoe*, *Moll Flanders*, and *Roxana*." *The Modern Language Review* 107, no. 1 (January 2012): 65–87.

Traver, John C. "Defoe, Unigenitus, and the 'Catholic' Crusoe." *Studies in English Literature, 1500–1900* 51, no. 3, Restoration and Eighteenth Century (Summer 2011): 545–563.

Tull, Jethro. *Horse Hoeing Husbandry*. London: G. Strahan in Cornhill; T. Woodward in Fleet-Street; A. Miller over-against St. Clement's-Church in the Strand; J. Stagg in Westminster-Hall; and J. Brindley in New-Bond-Street, 1731.

Van Sant, Ann. "Crusoe's Hands." *Eighteenth-Century Life* 32, no. 2 (Spring 2008): 120–137.

Walmsley, Peter. "Robinson Crusoe's Canoes." *Eighteenth-Century Life* 43, no. 1 (January 2019): 1–23.

Watt, Ian. *Rise of the Novel*. Berkeley: University of California Press, 1957. Kindle edition.

Wood, James Robert. "Robinson Crusoe and the Earthy Ground." *Eighteenth-Century Fiction* 32, no. 3 (Spring 2020): 381–406.

Zimmerman, Everett. "Robinson Crusoe and No Man's Land." *The Journal of English and Germanic Philology* 102, no. 4 (October 2003): 506–529.

3 Stewardship in American Literature
Promise and Paradox in the New World

Josh A. Weinstein

American nature writing and poetry of the natural world have grappled with the appropriate relationship of human beings to the natural world, arguably since their very inception. Here I will attempt to trace some of the key lines of development of one particularly widespread and, I would argue, especially promising approach to the natural world – that of stewardship, or care for the natural world.

Stewardship, at its root, may be defined as an attitude of care or custodianship that a responsible party – the steward, or stewards – exercises with regard to their charge, the person or entity entrusted into their care. This often takes the form of stewardship theology, or the belief that human beings have been called to fulfill a divine mission of care for the natural world, which in this view has been entrusted into human care by G-d.[1] This basic premise of stewardship theology, as Michael Northcott explains, holds

> that neither the Hebrew and Christian doctrine of creation, with its separation of creation from creator, nor the concept of human dominion over nature, involve a purely instrumentalist vision of nature which legitimates ecological plunder, because the role of humans with respect to nature is ordered by the metaphor of stewardship through which the Genesis creation accounts describe the human–nature relationship.[2]

While some may argue that this attitude or arrangement with respect to humans as stewards or caretakers of the earth, or nonhuman creation, is inherently anthropocentric, or human-centered, I would argue that the predominant formulation of an attitude of stewardship toward the natural world is actually profoundly *decentered* from the human concern, and is instead predicated upon human beings serving as humble intermediaries between the natural world and a higher power or system of relations that transcends humanity and is necessarily beyond our control, mastery or dominion whatsoever. In fact, the higher power or force that is understood to empower human beings to serve as stewards over nonhuman creation is so beyond us, whether that be G-d, or Gaia, or ecology, that efforts to understand or comprehend the animating force behind our role as stewards is typically described as futile, or necessarily partial and incomplete at best.[3]

In particular, I will attempt to show some similarities and confluences that exist between and amongst both religiously motivated versions of stewardship – such as the Christian stewardship theology pioneered by Susan Cooper in the mid-nineteenth century – and the scientifically motivated and scientifically informed stewardship promoted by authors such as Aldo Leopold and Rachel Carson in the middle of the twentieth century. We will go on to explore the perspective of contemporary ecopoet Gary Snyder, whose care for the natural world does not fit neatly, perhaps, into either a schema of religious or spiritually motivated stewardship, but instead melds aspects of each perspective in his attempts to make the world whole for us and for future generations through an attitude of care and wise, sustainable use of resources. Snyder, I will argue, exhibits a primary orientation that places human beings in humble relation to the world around us as a sort of caretaker for a world that is at once seen as beautiful and whole, as well as threatened by the depredations and greed of thoughtless overexploitation. This complexity and tension within Snyder's work, however, will ultimately lead us back to the complexity inherent in Cooper's work, over a century prior.

Susan Cooper and the Origins of Stewardship in American Literature

Susan Cooper, writing in the middle of the nineteenth century, makes an incredibly valuable contribution to the development of stewardship in American literature. Drawing on her Episcopalian Christian upbringing, Cooper articulates a vision of environmental stewardship theology that would ultimately become the predominant approach of religiously motivated environmentalists in the United States of America, and well beyond, both within the Episcopal Church, and across many other denominations. As Robert Booth Fowler explains:

> Since 1970, stewardship has been the environmentalist path most often proposed by Protestant Christians seeking to serve G-d and G-d's creations... Stewardship is encountered repeatedly in such liberal Protestant publications as the *United Methodist Reporter* or the *Presbyterian Survey*, but it has also been endorsed by the conservative General Council of the Assemblies of G-d, the Evangelical Lutherans, and the Episcopal Church, among many others.[4]

While some conservative Christians have pushed back on the environmentalist movement in recent years, and particularly with respect to attempts to address climate change, the predominant paradigm remains that of stewardship theology, or care for the natural world as G-d's creation.[5] It is nothing less than remarkable that Cooper, writing over a century earlier, and in the midst of the restless westward expansion of U.S. sovereignty over Native American lands, articulates so clearly her vision of a stewardship theology. This fact is even more striking when we consider Cooper's humility, not only before G-d, but also before her fellows in society and with respect to the land itself. Indeed, Cooper's attentiveness to the land of her home range during her years in Otsego County, New York, is indicative of a practice of humility toward the land which rejects human domination of the land and a rhetoric which subsumes nature. Rather, as Cooper explains in her introduction to *The Rhyme and Reason of Country Life*, human dominion of the earth is always already to be understood as a humble standpoint, for "even here on earth, within his own domains, his position is subordinate."[6] The interplay between Cooper's humility toward G-d, and the observations that her humble approach to nature enables lead her to achieve a more holistic view of the countryside than a mere naturalist admiring the world around her.

At first blush Cooper's *Rural Hours* seems to be pulled between the competing perspectives of mindful, attentive observations, and a more anthropocentric, Christian outlook, which is humble with respect to G-d, and humble toward nature only insofar as it represents G-d's creation.[7] Ultimately, however, it is Cooper's humility before her G-d that enables her to discover the ecological unity of the world in a way that does not compete with, but rather complements, her religious devotion. Lucy Maddox interprets Cooper's authorial role as fitting within the "more modestly feminine role of teacher and commonsensical interpreter of the moral lessons inherent in the natural processes of the rural place,"[8] whereby "simple and direct language" in describing nature serves as a primer to teach children about their American home environment,[9] thereby "reinforcing their attachment to their American rural homes."[10] I would argue, however, that to the extent that Cooper's descriptions are "simple and direct," this is a reflection of Cooper's practice of attentiveness toward the land, whereby simple language is the most faithful way of portraying nonhuman nature, and thus showing reverence for the divine force imbued in creation.

Cooper's use of the journal form in *Rural Hours* sets up her negotiation of these synergistic perspectives with respect to the place of humans and human communities in the natural world. As Tina Gianquitto argues, overtly Cooper seems to adopt a Christian natural theology of nature as the product of divine creation, for which humans are given the responsibility of wise stewardship.[11] When we consider other aspects of Cooper's texts, however, we find that her views of nature are more complex than this, and that often she sees nature as an ecologically interdependent network of relationships, whereby attentiveness to the minute details of the land, and the ecological humility this engenders, allows Cooper to intuit an essentially ecological ethics of care for the natural world that goes beyond a narrow notion of human stewardship to a sense of humans as being in community with the earth, even as they are granted a unique role in watching over and protecting it. Daniel Patterson, in his introduction to Johnson and Patterson's indispensable collection of Cooper's *Essays on Nature and Landscape*, argues that "For Cooper, religious faith is essential, though not sufficient alone [for the appreciation of nature]... It is only when one's knowledge is coupled with Christian faith that one can become capable of comprehending or representing nature properly and worthily."[12] Cooper's ability to intuit an ecological humility is remarkable not just for the depth and tension it engenders in her own perspective of religiously motivated stewardship of the earth, but also for the way in which it anticipates the complexity of contemporary ecological writers, such as ecopoet Gary Snyder, as we shall discuss below.

Patterson and others have identified Cooper's introduction to her 1854 anthology of nature poetry entitled *The Rhyme and Reason of Country Life* as a crucial exposition of her thoughts on religion and nature. The following passage, for example, marks out Cooper's vision of scriptural revelation as allowing humans to truly apprehend the natural world by "point[ing] out to man his own position, and that of all about him; he is lord of the earth and of all its creatures... all have been given into his hand – all are subject to his dominion."[13] This passage, taken out of the greater context of Cooper's introduction, and indeed that of her *Rural Hours* first published four years prior, would seem to clear the way for human beings to make use of nature in whatsoever way benefits us most. However, later in her introduction, Cooper makes clear that "while Revelation allots to him [Christians] a position elevated and ennobling, she also reads him the lowliest lessons,"[14] which is to say that Christian revelation for Cooper is inseparable from a deep feeling of humility before G-d and His creations. To have dominion over earthly creation is, for Cooper, still but a humble position compared to that of G-d who created all. For Cooper,

> No system connects man by more close and endearing ties, with the earth and all it holds, than Christianity, which... refers all to Providence, to the omniscient wisdom of a G-d who is love; but at the same time she warns him that he is himself *but the steward and priest of the Almighty Father, responsible for the use of every gift*; she plainly proclaims the fact, that *even here on earth, within his own domains, his position is subordinate. The highest relation of every created object is that which connects it with its Maker*[...][15]

The role of "steward and priest" of G-d's creation is a humbling one. According to Cooper, creation is noble in so far as it was created by G-d, and thus man's dominion over creation is not ennobling in itself, but rather as it serves to connect him that much more closely to his "Maker." As Johnson explains, this view represents

> what environmental ethicists now call 'stewardship' – a view that humans on earth must serve G-d even as they interact with the natural world, must recognize the natural world as G-d's creation, and must tend to nature on behalf of future generations.[16]

Christianity, for Cooper, encourages humans to reject a worldview that posits "man... [as] the great center of all" as hubristic,[17] and contrary to the "overpowering, heartfelt, individual humility" inspired by our apprehension of G-d's creation.[18]

Cooper's Christian humility posits human beings as an intermediary between G-d and His brute creations, neither abject, as with certain strains of Catholic humility, nor divine, except perhaps in our ability to fulfill a divine charge here on earth. Indeed, Cooper's statement in the same piece that "we need the view of the green turf, to teach us the humility of the grave" speaks back to a passage from the Jewish Talmudic text *Pirke Avot*,[19] whereby humans are called on to find humility in reflecting on our origin in the "fetid drop" of semen and the fact that we go at our death "to the place of dust, worms and maggots."[20]

We can find a similar emphasis on the relationship of death to a humble stance toward our lives as humans in the world in Cooper's entry for Monday, 11 September,[21] where she meditates upon the fading practice of using the rural churchyard as the burial place for local inhabitants:

> Nowhere is the stillness of the grave so deeply impressive; the feverish turmoil of the living, made up of pleasure, duty, labor, folly, sin, whirling in ceaseless movement about them, is less than the passing winds, and the drops of rain to the tenants of those grounds, as they lie side by side... The present, so full, so fearfully absorbing with the living, to the dead is a mystery; with those mouldering remains of man the past and the future are the great realities.[22]

Cooper thus invokes the "mouldering" bodies of the grave, literally returning as nutrients to the mould, or humus of the earth, in order to encourage this same idea of humility before the fact of our own mortality. The "ceaseless movement" of the business of our daily lives, that which distracts us from the possibility of mindful interactions with the world (in its social and more broadly ecological interactions) ought to be counterbalanced, Cooper argues here, with humility – an humility that recognizes both the emptiness of human endeavors for their own sake ("pleasure, duty, labor, folly, sin"), in an echo of the "all's vanity" message of Ecclesiastes, and the central importance of community, both with each other, and with the natural world. Thus, even after our deaths, we are destined to lie "mouldering" together into one soil, one humus, which will become the physical and figurative ground from which new life springs.

In Cooper's *Rural Hours* we find interspersed within her journal entries expressions of these views on Christian humility and the proper place of human beings in the world. For example, the following passage, from her entry for Tuesday, 16 May, represents Cooper's attempts to place her awe at the beauty of nature, here nature's springtime wonders, within a context of Christian humility before G-d's creation, complete with brief passages taken directly from scripture or paraphrased slightly:

> At hours like these, the immeasurable goodness, the infinite wisdom of our Heavenly Father, are displayed in so great a degree of condescending tenderness to unworthy, sinful man, as must appear quite incomprehensible – entirely incredible to reason alone – were it not for the recollection of the mercies of past years, the positive proofs of experience; while Faith, with the holy teaching of Revelation, proclaims "the L-rd, the L-rd G-d, merciful and gracious, long-suffering, and abundant in mercy and goodness." What have the best of us done to merit one such day in a lifetime of follies, and failings, and sins? The air we breathe so pure and balmy, the mottled heavens above so mild and kindly... all unite to remind us, that, despite our own unworthiness, "G-d's mercies are new every day."[23]

Here Cooper strikes a typical stance of Christian humility, reading through G-d's creation to find the hand of G-d at work, and in this particular case, even quoting and paraphrasing from scripture in the process. Cooper sees herself and all of humankind as fallen, and thus unworthy of G-d's grace and mercy. Her gratitude for this unmerited love from G-d is expressed through her humility toward the beautiful creation He has given humans to enjoy and care for as stewards. Further, it is striking here that Cooper merges this theological point with her concrete observations of nature. For Cooper, it is the "positive proofs of experience,"[24] along with faith in Christian revelation, rather than man's inferior reason, which allow humans to connect to G-d through the natural world. Gianquitto is correct to read an exegetic practice in Cooper's *Rural Hours*, as indeed Cooper at points seems to argue that our lives in nature and our experiences in the natural world represent our access to the "precious Book of Life... withdrawn from the cloisters and given to us all... alike full of the glory of Almighty Majesty, – great and worthy illuminations of the written Word of G-d."[25]

However, while Cooper does at times appropriate the language of Christian exegesis, and particularly the Puritan typological habit of reading in the natural world the word of G-d, she seems to exceed this model in significant ways. According to Christian biblical typology, the Old Testament provides the type for the New Testament antitype. By extension, the typological notion of the Two Books of G-d holds that scripture provides the type, to which the phenomenal world is but a negative impression, or antitype, an idea which many Puritan preachers and layman alike seized upon with zeal. As John Gatta explains, "Already in the Renaissance, the notion that G-d's sacred truths were inscribed in two books, that of the Bible and of Nature, had spread throughout European culture."[26] However, here Cooper refers to the "positive proofs of experience."[27] Rather than the world simply leading people back to scripture, as physical signs that represent biblical truths and enact scriptural designs, the world itself takes on the holiness of scripture – *the world not only corresponds to, but illuminates the words of scripture*. These "illuminations" in the natural world of creations,[28] as with the ornate calligraphic and pictorial illuminations of medieval religious texts, are linked to scripture, but go beyond mere correspondence to show the beauty of Providence at work in the natural world, and thus serve to inspire humans with awe and love for G-d. Cooper thus seems unable to limit the meaning of the natural world to a mere negative impression of positive scriptural meaning, but rather finds meaning in the "positive proofs of experience" that complement scripture.[29]

This mindful stance is further evidenced in Cooper's meditation on the connections between species native to her central New York home of Otsego County, and similar species in far-flung parts of the world. For example, in her entry for Tuesday, 29 May, she notes that the species of "mandrakes, or May-apples" flowering in her area "is said also to be found under a different variety in the hilly countries of Central Asia," ultimately leading her to one of her most eloquent pronouncements of ecological interconnection in *Rural Hours*: "One likes to trace these links," she writes, "connecting lands and races, so far apart, reminding us, as they do, that the earth is the common home of all."[30] This is, in essence, the foundational intuition of ecology as a science and the basis of an holistic world-view – an attentiveness to the interconnections of all of the various parts that make up our whole home systems, the study of our home (*oikos*, eco-) as one system.

Richard Magee describes Cooper's work as laying out a "sub-genre... [of] sentimental ecology, whereby the demands of community and domestic life are intertwined, much like models of ecosystems, with the demands of the natural environment."[31] While I am uncertain that this sub-genre holds up in relation to the often-unsentimental accounts of nature Cooper provides in *Rural Hours*, Magee is certainly correct in drawing attention to Cooper's concern with the interpenetration of the domestic and the natural. Indeed, if "the earth is the common home of all,"[32] as Cooper declares, then a true ecological vision is one which takes into account the multiple overlapping and interpenetrating levels of home and household of which we are each a part.

Thus, Cooper's humble ecological vision and her Christian beliefs prove to be mutually supporting. Just as her minute mindfulness of the land leads her to understand sophisticated interconnections between various aspects of her human and nonhuman environment in ways which we clearly recognize as ecological, these same observations also lead her back to her G-d, through expressions of gratitude for the bounty of Providence. While idiosyncratic from a standpoint of strict philosophical or theological inquiry, Cooper, as a complex, human figure, is peculiarly adept at keeping these different ways of understanding her world in play simultaneously. Thus, it is not simply a matter of conflicting systems of thought, but rather a tension between two systems of thought which sometimes manifests itself as a seeming oppositional pull or twist to be reconciled, and at other times as a synergistic urge toward a view that accomplishes a harmonious merging of these differing perspectives.

One such moment of productive tension in Cooper's *Rural Hours* is her description and meditation on the plentiful autumn nuts, dated November 1:

> The beech-trees are dotted with nuts. The wych-hazel has opened its husks, and the yellow flowers are dropping with the ripe nuts from the branches. Acorns and chestnuts are plentifully scattered beneath the tress which bore them. How much fruit of this sort, the natural fruit of the earth – nuts and berries – is wasted every year; or rather, how bountiful is the supply provided for the living creatures who need such food![33]

Here Cooper seems to pivot between a view of the prodigality of nature as "waste," based upon the implicit measure of how much human beings are able to gather and make use of, and an ecological vision which realizes that what is excess with respect to the direct, immediate needs and desires of human beings represents a necessary food source for other living creatures. In other words, Cooper is able to reflect, even if briefly, upon her perspective as a member of a human society, as well as an alternative perspective as part of an ecosystem in which multiple organisms may partake of a given food source, either competitively or harmoniously, as in this example.

For Cooper, then, the interaction of human beings with the land is both an opportunity to exercise wise stewardship of the earth as natural resources to be caringly used in an effort to ensure what we now call "sustainable development," or a sense of "farther progress," as Jennifer Dawes argues (155),[34] as well as an occasion to develop a deeper sense of love and awe of G-d through not only contemplating, but striving to understand the complexities of created nature. Although Cooper is not wholly committed to a nascent ecological sense of the interconnected "world as home" as the sole paradigm by which to reconfigure and restructure our interactions with the land, by offering us the concept of the earth as creation entrusted to human care by G-d, alongside her insight of the earth as "the common home of all,"[35] an intuition which forms the cornerstone of our modern science of ecology, she proves to be an early proponent of a now mainstream Christian doctrine of Christian stewardship of the natural world.[36]

The way Cooper's *Rural Hours* negotiates this central relationship between Christian stewardship theology and ecological humility toward the ecosystems of the earth themselves may, therefore, prove useful in helping to frame and further current environmental debates, helping to tip the balance toward not just a "greener" America, but a "greener" Earth. It seems that in order to remain focused on this noble end we must be constantly reminded, as Cooper was, "that the earth is the common home of all."[37]

Aldo Leopold, Rachel Carson, and the Science of Environmental Stewardship

One of the most fascinating, and promising, aspects of stewardship though as a practical approach to safeguard ecosystems from harm is that, while it may be animated by religious feeling and devotion to an all-powerful G-d above – as I argue above is the case for Susan Cooper – it need not be confined to those professing religious feelings whatsoever. We find two prime examples of a scientific approach to stewardship in Aldo Leopold's *A Sand County Almanac and Sketches Here and There* (1949),[38] and Rachel Carson's seminal work *Silent Spring* (1962).[39] In these two works, we find a vision of stewardship that is independent from religious and spiritual fervor. Rather, in the works of these two authors, there arises a convergent approach of stewardship of the natural world, as it were, on the basis of scientific evidence pointing to the utility of a humble, custodial approach to the natural world.

First, let us look at Leopold's version of stewardship, as reflected in *A Sand County Almanac*, and in the essays "Thinking Like a Mountain" and "The Land Ethic" in particular. In "Thinking Like a Mountain," an anecdote of Leopold's personal experience as a young sport hunter, and his subsequent retrospective realizations from the perspective of a more mature and ecologically well-informed Leopold. As a young man, he reflects,

> I thought that because fewer wolves meant more deer, that no wolves would mean hunters' paradise. But after seeing the green fire [in the old wolf's eyes] die, I sensed that neither the wolf nor the mountain agreed with such a view.[40]

Only later on, after he "live[s] to see state after state extirpate its wolves," now armed with the knowledge and experiences of being a pioneer in the ecological management of forests does Leopold come to understand what it is that he began to sense on that fateful day on the mountain: that there is a thought-process inherent to the land itself, here "the mountain," that transcends the relatively petty and short-sighted concerns of human beings and sees to the interests of the ecosystem as a whole. To "think like a mountain," then, is to do right by the land, where "A thing is right when it tends to preserve the integrity, stability, and beauty of the biotic [living] community. It is wrong when it tends otherwise,"[41] as Leopold explains in the conclusion of his essay "The Land Ethic."

In fact, his comment in the brief vignette that is "Thinking Like a Mountain" comparing a mountain after its wolves have been extirpated as "look[ing] as if someone had given G-d a new pruning shears, and forbidden Him all other exercise" can be understood as implicitly acknowledging the outsized importance and ecological impact of human actions on the land community.[42] After all, in this example, were it not for human hunters and range managers deciding to eliminate wolves from the landscape, there would be no drastic "pruning" of the mountain's vegetation from the resulting overpopulation of deer. Whether or not we accept the role of custodian and caretaker of the land community, it is impossible for Leopold the scientist to ignore the empirical evidence of our ability as humans to shape the land with which we interact.

In Leopold's famous essay "The Land Ethic," he approaches the topic of where human beings fit into ecosystems even more explicitly. At first blush, Leopold's orientation here may seem to have little to do with stewardship. After all, he describes the relationship of humans to other organisms as being essentially on a level playing field, ethically speaking, as in the following passage summing up his main argument in the essay: "In short, a land ethic changes the role of *Homo sapiens* from conqueror of the land-community to plain member and citizen of it. It implies respect for his fellow members, and also respect for the community as such."[43] In other words, human beings are not to be considered as dominant over the rest of the land community and are perhaps not even to be regarded as stewards of nonhuman creation according to Leopold's formulation here. Rather, we are simply "fellow-members" of the "land-community" – and "plain member[s]" at that! The implication, then, is that humans have no special privileges or responsibilities allotted to us with respect to the other members of the ecosystems that sustain us.

Similarly, Leopold describes the humble station of humans' situation further in "The Land Ethic," when explaining the humility of the ecological scientist in interpreting human history, and our tendency – especially among the scientifically less-educated – to overestimate our knowledge of how the land works and to what uses it ought to be put. There Leopold writes,

> man is in fact only a member of a biotic team… Many historical events, hitherto explained solely in terms of human enterprise, were

actually biotic interactions between people and land. The characteristics of the land determined the facts quite as potently as the characteristics of the men who lived on it.[44]

However, despite Leopold's declaration of our plain status as members of the land community, and the humbling realization of our dynamic interactions with the land, if we dig a bit deeper here, we will see that Leopold *does* in fact have a special destiny in mind for human beings. His vision for humans is one that attempts to leverage and draw upon our distinct capacities both to have an outsized impact on the land, as well as our ability to study, measure and alter this impact.

For example, further on in "The Land Ethic," in his effort to promote an ecological conscience that recognizes the rights of other members of the biotic community irrespective of their economic use-value, Leopold writes: "If the private [land] owner were ecologically minded, he would be proud to be the *custodian* of a reasonable proportion of such areas [that lack economic value], which add diversity and beauty to his farm and to his community."[45] While Leopold largely avoids speaking of humans as having been especially called to care for nonhuman creation, moments such as the one quoted above betray his own underlying orientation that the land ethic he advocates is predicated upon our willingness to serve as "custodian[s]" of the land, and not merely as plain members of the land community. After all, if the central premise of Leopold's land ethic is that we must accept an "ethical obligation toward land,"[46] and that it is essential for us to see beyond our own narrow self-interest in the pursuit of healthy ecosystems, this can only occur, it seems, by our realizing not only that we depend upon the land, but that to a degree not fully explainable by our biology, the land depends on our wise actions, that is our active *stewardship* of the land.

Similarly, in the case of Rachel Carson's *Silent Spring*, we find Carson advocating for a wise management, or stewardship, of the earth and its living communities based on the outsized effect of human actions and specifically our use of chemical insecticides in "man's war against nature."[47] For example, in arguing against the indiscriminate spraying of chemical insecticides, or "biocides" as she would call them,[48] on suburban lawns, she cites "[t]he success of selective spraying for roadside and right-of-way vegetation" as an example of "sound ecological methods [… used in] *managing* vegetation as a living community."[49] The very language here of "managing vegetation," I would argue, betrays Carson's deeply held view that human beings bear a responsibility to care for the earth as wise stewards, or custodians, of the complex living communities we inhabit.[50]

One key difference, though, between the scientifically motivated, and more specifically ecologically motivated stewardship of Leopold and Carson and the religiously-inspired stewardship of Cooper, is that the status of human beings as custodians and managers of ecosystems in the scientifically inspired schema is typically seen as inferior to the inbuilt capacities of ecosystems – or "Nature herself"[51] – to self-regulate and achieve ecosystem balance, harmony, or homeostasis. For example, in the same section of *Silent Spring* referenced above, Carson goes on to opine that

> Nature herself has met many of the problems that now beset us, and she has usually solved them in her own successful way. Where man has been intelligent enough to observe and to emulate Nature, he too is often rewarded with success.[52]

In other words, while serving as de facto stewards of the land, Carson, and indeed Leopold, seem to view human stewardship of the land as second-best to the natural world's own inherent mechanisms for achieving and maintaining a dynamic ecological equilibrium.

Indeed, in the case of Carson in particular, the humility that Cooper displays before G-d (as "but the steward and priest of the Almighty Father"[53]) seems to be replaced by an ecological humility, oriented toward "Nature herself."[54] This shift from serving as stewards of G-d, to serving as de facto stewards of living systems that are beyond our understanding, or even as a system animated by an entity other than G-d (such as "Gaia,") is a crucial aspect of the paradox of convergence in views on stewardship of the earth. Bernard Zaleha and Andrew Szasz, for example, argue that "conservative Christians," and evangelicals in particular, harbor "suspicions that 'stewardship,' improperly understood, smacks of neo-pagan-style nature worship."[55]

Furthermore, as with Leopold and his land ethic, Carson is animated by a desire to live as an ethically upright human being, rather than to serve as a "steward and priest" of G-d, as with Cooper. For example, at the conclusion of her chapter "Needless Havoc," which decries the indiscriminate overapplication of chemical insecticides and the ill effects these practices engender, Carson closes the chapter with the following rhetorical question: "By acquiescing in an act that can cause such suffering to a living creature, who among us is not diminished as a human being?"[56] In other words, by causing the unintended suffering and death of animal species in our "war against nature"[57] – here the gruesome poisoning of a ground squirrel – we are harming our own humanity.

Interestingly, whether the motivation for stewardship of the land is humility before the supreme G-d, or an ecological humility before the unknown – and perhaps unknowable – complexity and intricacy of the ecosystems of the world, both orientations lend themselves to a similar appreciation for the responsibilities we have as human beings to care for the natural world. A key challenge in achieving practical synergy in addressing environmental challenges between adherents of these distinct, yet convergent viewpoints, may be figuring out how to maintain focus on the common ground of a belief in caring for the natural world, while fostering an openness to a diversity of opinions on the first principles that motivate disparate groups to do so.[58]

Gary Snyder's *Turtle Island* and the Ecopoetics of Stewardship

Contemporary ecopoet and essayist Gary Snyder, in his own way, nicely demonstrates a complex interplay of both aforementioned main strains of stewardship – scientifically motivated on the one hand and religiously or spiritually motivated on the other. In Snyder's 1974 book of poems, *Turtle Island*,[59] published to include his essay "Four Changes," we see the work of a visionary poet and radical thinker attempting to effect a call to action that tends toward a fundamental reordering of the role human beings play in the world with respect to other organisms and ecosystems and the ecosphere as a whole, or what Snyder calls "transformation."[60] From the very start of "Four Changes," Snyder calls on humankind to "act as *gentle steward* of the earth's community of being."[61] While Snyder's views here are clearly informed by an understanding of the science of ecology, and akin to those of Leopold or Carson in this respect, Snyder's views go beyond a scientific view of our place in the natural world to engage with the cultural, religious, and spiritual dimensions of our interconnected coexistence with earth.

In "Four Changes," Snyder advocates for key changes in the following four areas: "population," "pollution," "consumption" and "transformation." The first three categories are, perhaps, self-explanatory, and call, in part, for (1) a drastic reduction in world population to "half of the present [c. 1969] world population or less,"[62] (2) "Clean air, clean clear-running rivers,"[63] and (3) a decrease in consumption such that we achieve "Balance, harmony, humility, growth which is a mutual growth with Redwood and Quail."[64] The fourth category of "transformation,"[65] however, is perhaps the most challenging, in that it calls for us to "change the very foundations of our society and our minds" to achieve "a planet on which the human population lives harmoniously and dynamically by employing various sophisticated and unobtrusive technologies in a world environment which is 'left natural.'"[66] Key to this vision of the role of human beings on Planet Earth is a two-fold transformation. On the one hand, Snyder encourages the exploration of any number of traditions of thought and practice that he lists under the rubric of "hav[ing] worked through history toward an ecologically and culturally enlightened state of affairs."[67] This "revolution of consciousness" for Snyder,[68] however, must not supplant the hard work of making actual change in the (physical) world. As Snyder writes parenthetically:

> More concretely: no transformation without our feet on the ground. *Stewardship* means, for most of us, find your place on the planet, dig

in, and *take responsibility* from there – the tiresome but tangible work of school boards, county supervisors, local foresters – local politics.[69]

The key for Snyder, then, is to take responsibility for the particular places we call home, even as we recognize that ecological concern is ultimately planetary in scope. We must work in our own communities "[e]ven while holding in mind the largest scale of potential change" – Snyder's articulation, as it were, of the famous environmentalist slogan: "think globally, act locally."

Snyder's essay here builds on citizen-scientist insights, such as an understanding of the ecological concept of *carrying capacity*, or the total number of organisms of a particular kind that an ecosystem can support. Snyder's views here, however, build on this concept to include the well-being of the entire planetary ecosystem as follows: "the fact stands that by standards of planetary biological welfare there are already too many human beings."[70] Here, instead of focusing of the narrow concept of a species-specific carrying capacity, Snyder arrives at a goal for a maximum human population number "based on the sense of total ecological health for the region [in question], including flourishing wildlife populations."[71] As with Carson's science-based critique of the use of chemical pesticides and other risky behavior toward the natural world, even as Snyder leverages insights from the science of ecology, he acknowledges the overarching importance of human actions relative to the environment and nonhuman life forms, and seems haunted by the specter of destruction that humans, and only humans, are capable of inflicting on the natural world, and ultimately upon ourselves.

This anxiety is apparent in Snyder's poem "L M F B R," which appears toward the end of the second poetic section of the book: "Magpie's Song," just a couple of poems before the section ends and the last section of poems in *Turtle Island*, "For the Children," begins. An acronym for "Liquid Metal Fast Breeder Reactor" that the poet spells out for us in the second line of the poem,[72] this poem seems to veritably ooze and glow with death, destruction and the perversion of the natural world as wrought by a headlong race to master nuclear power generation.

While "L M F B R" predates major nuclear power accidents such as Three Mile Island, Pennsylvania (1979), or Chernobyl, Ukraine (1986), Snyder channels the already nascent environmentalist-driven anti-nuclear power movement – personifying human energy dependence as a grim reaper of sorts: "Death himself."[73] As he writes in "Energy is Eternal Delight," a short essay that follows "Four Changes" in the "Plain Talk" section of *Turtle Island*, Snyder is convinced that the major industrial economies of the world are "addicted to heavy energy use," and that "As fossil-fuel reserves go down, they will take dangerous gambles with the future health of the biosphere (through nuclear power) to keep up their habit."[74] Here, the opposition to nuclear power is joined seamlessly to the rejection of other exploitative and wasteful human actions. Thus, Death-as-energy dependence grins with "Plutonium tooth-glow" and is clutching a "Strip-mining scythe,"[75] symbolizing that the extraction of natural resources, such as coal, through strip-mining is akin to bringing death upon our natural environment. Again, Snyder's "plain talk" in "Energy is Eternal Delight" is instructive here, as he references the "cancer [that] is eating away at the breast of Mother Earth in the form of strip-mining," with specific mention of the Black Mesa coal mine that is "[o]n Hopi and Navajo land... sacred territory."[76] Similarly, Snyder lists several man-made, non-biodegradable materials – such as "plastic spoons" and "PVC pipe" that "don't exactly burn, don't quite rot, / flood over us" as we approach the "end of days."[77] The disturbing vision in "L M F B R," then, is one of radical disharmony with the natural world that ultimately threatens human life on Earth, as well as damage to the health of the biosphere, or the earth understood as one vast habitat for life.

There is a similar logic of environmental destruction conveyed in the poem "Mother Earth: Her Whales," which appears toward the middle of the "Magpie's Song" section of *Turtle Island*. Here we find the beauty of wild creatures such as the titular whales that "turn and glisten, plunge / and sound and rise again,"[78] juxtaposed with destructive human actions such as rainforest destruction ("The living actual people of the jungle / sold and tortured"[79]) and ocean water pollution (Japan "dribbles methyl mercury / like gonorrhea / in the sea"[80]). However, when we consider both of these poems in the context of "Four Changes," it becomes apparent that there is a paradox of sorts. On the one hand, humankind, "As the most highly developed tool-using animal," has a special responsibility to "act as gentle steward of the earth's community of being."[81] On the other hand, it becomes clear that this role of stewardship is not mainly for the benefit of the earth, but rather that of human beings ourselves.

In fact, the closest Snyder comes in *Turtle Island* to describing an apocalypse of sorts comes with an enigmatic series of three unattributed quotes just shy of the conclusion of "Four Changes" where Snyder, drawing on Buddhist cosmology, writes: "No need to survive." "In the fires that destroy the universe at the end of the kalpa, what survives?" – "The iron tree blooms in the void."[82] While a full consideration of the Buddhist concept of kalpa – or aeons – is beyond the scope of this essay, it seems clear that this statement is meant to frame human action within the humbling scope of aeons of time, that are by definition mind-bogglingly long (16 million years? 1 billion years?), rather than present the specter of a fate that may be brought on or staved off by human actions. Indeed, in a 1988 interview with Julia Martin, Snyder explains the line that follows those cited above to end the essay – "Knowing that nothing need be done, is where we begin to move from" – as follows: "In the larger scale, things will take care of themselves. It's obviously human hubris to think we can destroy the planet, can destroy life. It's just another exaggeration of ourselves. Actually we can't. We're far too small."[83]

Perhaps a more accurate vision of Snyder's hopes and fears for humanity, then, is to be found in the final section of poetry in *Turtle Island*, "For The Children," and in particular in the poem of that same title. The first stanza of the poem sketches out the problem of overpopulation and approaching overshoot of the human carrying capacity of the earth. He describes a figurative "steep climb" on "rising hills, the slopes / of statistics," followed by the figurative descent of humanity – "everything going up / up" (i.e. population, resource exploitation), "as we all / go down" – presumably our quality of life and our ability to support ourselves on the land indefinitely and without causing lasting harm.[84] That is, our ability to live in an ecologically wholesome manner is diminished as our population continues to rise and approach the earth's carrying capacity.

The second stanza of "For The Children" leads us into a vision of the future. The "steep climb" of statistics from stanza one, yields to imagery of "valleys, pastures," again overlaying the rise and fall of statistics with that of topography, only here we encounter the flattening of statistical graph lines melding with a topography that is more amenable to supporting human life, including agricultural activity and hunter-gatherer ways.[85] Published in 1974, in the midst of the U.S.-Soviet Union Cold War, it is also noteworthy that this future is described as a place in space & time where "we can meet there in peace / if we make it."[86] The implication here, then, is that the main danger we face as a species is one of self-harm through our shortsighted and unwise use of the natural world. The fourth and final stanza of the poem crystallizes the lessons to be learned as follows:

> *stay together*
> *learn the flowers*
> *go light*[87]

We can explicate these instructions approximately as follows: *stay together* – practice unity and solidarity with one another and other lifeforms; *learn the flowers* – learn (and pass on) ethnobotanical knowledge; *go light* – simplify, reject the need for continual growth.

In the final analysis, then, Snyder's view on stewardship here, I would argue, evinces both a scientifically motivated and a spiritually motivated stewardship. Snyder's outlook is similar to that of Carson's in *Silent Spring* and Leopold's in *A Sand County Almanac* in that it is informed by science, and skeptical of human ability to meet the challenges of fostering and maintaining a healthy relationship with the natural world. It is also, however, based on a related spiritual insight of the radical interdependence of all lifeforms, and of all life on Earth, with the very earth itself. For Snyder, the earth is not simply a "land-community," as it is for Leopold, but rather a

> great
> earth
> sangha[88]

Borrowing a Buddhist term for a community of religious leaders and lay people, Snyder suggests that we are all members in a vast spiritual community of the entire earth, interconnected and interdependent, despite our various cultures, traditions, and stories. From this perspective, then, the best kind of stewardship – healthiest for humankind and for the biosphere at large – is that which involves human beings simultaneously relinquishing our outsized role in the workings of the natural world and joining cooperatively in a great spiritual community with the ecosystems, of which we form but a small part.

What we can learn, then, from these various case studies of stewardship in American literature is both the diversity of this perspective, as well as its durability and flexibility. Stewardship need not be linked to religious fervor or scriptural basis, but even when it is, such as for Cooper, there is still room for complexity and tension within these views and beliefs. Similarly, stewardship need not be wholly based on scientific insight either – as it seems to be for Leopold and Carson – but rather it can be based on a melding of spiritual and scientific orientations, as with Snyder.

This diversity and flexibility, I would argue, holds great promise for humankind's present-day struggle with ecological challenges, and in particular the struggle to adequately address human-induced climate change. The central insight and call to action of stewardship – that human beings must take responsibility to care for the earth and its lifeforms and habitats, for our own health and the health of the planet – is fundamental to the type of broad-based societal and economic change that is needed to meet our current challenges. Literature, and specifically the literature of stewardship, has played a critical role in popularizing and disseminating key ecological insights in the past, and will no doubt continue to play a crucial role in the fight against climate change and other ecological challenges.[89]

Notes

1 In accordance with Jewish custom, I have omitted the letter -o- here in reference to G-d. In the following pages and source quotations, unless otherwise noted, the omission is my own.
2 Michael S. Northcott. *The Environment and Christian Ethics* (Cambridge: Cambridge University Press, 1996), 129. Northcott here is specifically referencing Robin Attfield's defense of stewardship. For more on stewardship, see

Roger Gottlieb, *A Greener Faith* (Oxford: Oxford University Press, 2006), especially regarding the apparent conflict between the tendency toward both "environmental stewardship and unrestrained economic growth" of "many nineteenth-century Protestant industrialists" (30), which contrasts with Susan Cooper's more thoughtful approach toward achieving harmony between human activities and the natural world discussed below.

3 For more on the Gaia hypothesis in particular, see James Lovelock, *Gaia: A New Look at Life on Earth* (Oxford: Oxford University Press, 2000). For Lovelock's critique of the limitations of environmental stewardship, see James Lovelock, "The Fallible Concept of Stewardship of the Earth," in *Environmental Stewardship: Critical Perspectives – Past and Present*, ed. R. J. Berry (London: T&T Clark International, 2006), 106–11.

4 Robert Booth Fowler. *The Greening of Protestant Thought* (Chapel Hill: University of North Carolina Press, 1995), 76.

5 See, for example, Bernard Daley Zaleha and Andrew Szasz, "Why Conservative Christians Don't Believe in Climate Change," *Bulletin of the Atomic Scientists* 71, no. 5 (2015): 19–30, which argues that "[c]onservative Christians... constitute one of the core demographic foundations for climate skepticism" (20). See also, Pew Research Center's 2015 report, "Religion and Views on Climate and Energy Issues," which notes that "[w]hite evangelical Protestants stand out as least likely" to view global warming as "due to human activity." However, the same report claims that: "Political party identification and race and ethnicity are stronger predictors of views about climate change beliefs than are religious identity or observance," https://www.pewresearch.org/science/2015/10/22/religion-and-views-on-climate-and-energy-issues/.

6 Susan Fenimore Cooper, *The Rhyme and Reason of Country Life* (New York: Putnam, 1855), 28.

7 Susan Fenimore Cooper, *Rural Hours*, eds. Rochelle Johnson and Daniel Patterson (Athens, GA: University of Georgia Press, 1998).

8 Lucy Maddox, "Susan Fenimore Cooper's Rustic Primer," in *Susan Fenimore Cooper: New Essays on Rural Hours and Other Works*, eds. Rochelle Johnson and Daniel Patterson (Athens, GA: University of Georgia Press, 2001), 83–96, 88.

9 Maddox, "Susan Fenimore Cooper's Rustic Primer," 94.
10 Maddox, "Susan Fenimore Cooper's Rustic Primer," 94.
11 Tina Gianquitto, "The Noble Designs of Nature: G-d, Science, and the Picturesque in Susan Fenimore Cooper's *Rural Hours*," in *Susan Fenimore Cooper: New Essays on Rural Hours and Other Works*, eds. Rochelle Johnson and Daniel Patterson (Athens, GA: University of Georgia Press, 2001), 169–90.

12 Daniel Patterson, "Introduction," in Susan Fenimore Cooper, *Essays on Nature and Landscape*, eds. Rochelle Johnson and Daniel Patterson (Athens, GA: University of Georgia Press, 2002), xviii.

13 Cooper, *Rhyme and Reason*, 22.
14 Cooper, *Rhyme and Reason*, 28.
15 Cooper, *Rhyme and Reason*, 28. My emphasis.
16 Rochelle Johnson, "Placing *Rural Hours*," in *Reading under the Sign of Nature: New Essays in Ecocriticism*, eds. by John Tallmadge and Henry Harrington (Salt Lake City, UT: University of Utah Press, 2000), 77.
17 Cooper, *Rhyme and Reason*, 28.
18 Cooper, *Rhyme and Reason*, 29.
19 Cooper, *Rhyme and Reason*, 30.
20 Avot 3.1, in *The Ethics of the Talmud: Sayings of the Fathers*, ed. and trans. R. Travers Herford (New York: Schocken Books, 1971), 63.

21 While Cooper does not include the year for her entries in *Rural Hours*, based on her preface to the book and the entries that include the day of the week, we can infer that she is primarily recording her observations from March 1848 through February 1849. As Cooper historian Hugh C. MacDougall points out in an online annotation to an early article on *Rural Hours*, however, she seems to compensate for some travel away from Cooperstown in the summer of 1848 by "shifting to [observations from] 1849 for much of the summer months... and then reverting to 1848 and following through to February 1849" (note 3a to online reprint of: Anna K. Cunningham, "Susan Fenimore Cooper – Child Genius" [1944], via the James Fenimore Cooper Society website, https://jfcoopersociety.org/susan/1944NYHistory-Cunningham.html).
22 Cooper, *Rural Hours*, 180.
23 Cooper, *Rural Hours*, 45. The first quote here is from Exod. 34:6. The latter is an approximation of parts of the following verses: Lam. 3:22–23.
24 Cooper, *Rural Hours*, 45.
25 Cooper, *Rural Hours*, 46.
26 John Gatta, *Making Nature Sacred: Literature, Religion, and Environment in America from the Puritans to the Present* (Oxford: Oxford University Press, 2004), 73.
27 Cooper, *Rural Hours*, 45.
28 Cooper, *Rural Hours*, 46.
29 Cooper, *Rural Hours*, 45.
30 Cooper, *Rural Hours*, 56.
31 Richard M. Magee., "Sentimental Ecology: Susan Fenimore Cooper's *Rural Hours*," in *Such News of the Land: US Women Nature Writers*, eds. Thomas S. Edwards and Elizabeth A. De Wolfe (Hanover, New Hampshire: University Press of New England, 2001), 28.
32 Cooper, *Rural Hours*, 56.
33 Cooper, *Rural Hours*, 229.
34 Jennifer Dawes. "A Farther Progress: Reconciling Visions of Time in Susan Fenimore Cooper's *Rural Hours*," in *Susan Fenimore Cooper: New Essays on Rural Hours and Other Works*, eds. Rochelle Johnson and Daniel Patterson (Athens, GA: University of Georgia Press, 2001), 154–168, 155.
35 Cooper, *Rural Hours*, 56.
36 It is interesting to note that even those religious groups that Zaleha and Szasz identify as most antagonistic to calls for climate change action, such as "the Cornwall Alliance for the Stewardship of Creation," nevertheless maintain the concept of "Biblical earth stewardship" as central to their faith-based approach to the earth (Zaleha and Szasz, "Why Conservative Christians Don't Believe in Climate Change," 26; "Cornwall Alliance for the Stewardship of Creation: About," https://cornwallalliance.org/about/.)
37 Cooper, *Rural Hours*, 56. For a fuller discussion of Susan Fenimore Cooper's ideas on Christian stewardship and her way of finding connection to the land, see also: Josh A. Weinstein, "Susan Cooper's Humble Ecology: Humility and Christian Stewardship in Rural Hours" (*Journal of American Culture* 35, no. 1, 2012), 65–77.
38 Aldo Leopold, *A Sand County Almanac and Sketches Here and There* (New York: Oxford University Press, 1949).
39 Rachel Carson, *Silent Spring* (New York: Mariner Books, 2002).
40 Leopold, *Sand County Almanac*, 130.
41 Leopold, *Sand County Almanac*, 224–25.
42 Leopold, *Sand County Almanac*, 130–32.

43 Leopold, *Sand County Almanac*, 204.
44 Leopold, *Sand County Almanac*, 205.
45 Leopold, *Sand County Almanac*, 212. My emphasis.
46 Leopold, *Sand County Almanac*, 214.
47 Carson, *Silent Spring*, 7.
48 Carson, *Silent Spring*, 8.
49 Carson, *Silent Spring*, 81. My emphasis.
50 It is interesting to note that Carson does not use the term *stewardship* in *Silent Spring*, and only twice uses the title "custodian," both times in reference to a specific person in charge of a particular delineated bird sanctuary. This lexical omission notwithstanding, I argue here that Carson's vision of human involvement in the wise management of the land fits squarely within the bounds of stewardship as we are exploring the idea here. Indeed, as Carson's biographer Linda Lear succinctly summarizes Carson's main call to action in *Silent Spring* as follows: "She calls for humans to act responsibly, carefully, and as *stewards* of the living earth" ("Silent Spring," rachelcarson.org/SilentSpring.aspx, my emphasis).
51 Carson, *Silent Spring*, 81.
52 Carson, *Silent Spring*, 81.
53 Cooper, *Rhyme and Reason*, 28.
54 Carson, *Silent Spring*, 81.
55 Zaleha and Szasz, "Why Conservative Christians Don't Believe in Climate Change," 20.
56 Carson, *Silent Spring*, 100.
57 Carson, *Silent Spring*, 7.
58 Other challenges include avoiding the politicization of environmental issues and gaining consensus on complex environmental problems. One takeaway from the current debate about addressing climate change is the extreme difficulty in gaining consensus on the need for dramatic change to address global warming, when the steady rise of carbon dioxide and other greenhouse gases in our atmosphere is impossible to perceive first-hand with our human senses, but must be investigated indirectly, via scientific observations etc. Unlike water and air pollution, which are often perceptible outright, and cause clear and harmful effects (foul-smelling water, acrid air, etc.), complex phenomena such as global warming and human-induced climate change are more difficult to understand, and the attribution of human actions as causative is more difficult to clearly establish. The current trend of increasing intensity and frequency of natural disasters such as hurricanes, forest fires, and flooding, and the increasingly clear connection to human-induced global warming, however, may be a tipping point in achieving consensus and taking decisive action on the level of whole societies, and indeed on a global basis.
59 Gary Snyder, *Turtle Island* (New York: New Directions, 1974).
60 Snyder, *Turtle Island*, 99.
61 Snyder, *Turtle Island*, 91.
62 Snyder, *Turtle Island*, 92.
63 Snyder, *Turtle Island*, 94.
64 Snyder, *Turtle Island*, 97.
65 Snyder, *Turtle Island*, 99.
66 Snyder, *Turtle Island*, 99–100.
67 Snyder, *Turtle Island*, 100.
68 Snyder, *Turtle Island*, 101.
69 Snyder, *Turtle Island*, 101. My emphasis.
70 Snyder, *Turtle Island*, 92.

70 Josh A. Weinstein

71 Snyder, *Turtle Island*, 93.
72 Snyder, *Turtle Island*, 67.
73 Snyder, *Turtle Island*.
74 Snyder, *Turtle Island*, 103.
75 Snyder, *Turtle Island*, 67.
76 Snyder, *Turtle Island*, 104.
77 Snyder, *Turtle Island*, 67.
78 Snyder, *Turtle Island*, 47.
79 Snyder, *Turtle Island*.
80 Snyder, *Turtle Island*.
81 Snyder, *Turtle Island*, 91.
82 Snyder, *Turtle Island*, 102.
83 Gary Snyder, "Nothing Need Be Done," interview by Julia Martin, *Tricycle*, Spring 2015, https://tricycle.org/magazine/nothing-need-be-done. Adapted from Julia Martin and Gary Snyder, *Nobody Home: Writing, Buddhism, and Living in Places* (San Antonio, TX: Trinity University Press, 2014).
84 Snyder, *Turtle Island*, 86.
85 Snyder, *Turtle Island*.
86 Snyder, *Turtle Island*.
87 Snyder, *Turtle Island*. Original emphasis.
88 Snyder, *Turtle Island*, 73.
89 While an author such as Snyder, whose writings are deeply influenced by Zen Buddhism, is unlikely to be an influential voice with evangelical Christian readers, his friend and popular Christian ecopoet Wendell Berry is well-positioned to do so. See, for example, Wendell Berry, *A Timbered Choir: The Sabbath Poems 1979–1997* (Washington, DC: Counterpoint, 1998).

Bibliography

Carson, Rachel. *Silent Spring*. 1962. New York: Mariner Books, 2002.
Cooper, Susan Fenimore. *Rural Hours*. 1850. Edited by Rochelle Johnson and Daniel Patterson. Athens, GA: University of Georgia Press, 1998.
Cooper, Susan Fenimore. *The Rhyme and Reason of Country Life*. New York: Putnam, 1855. 13–34.
Dawes, Jennifer. "A Farther Progress: Reconciling Visions of Time in Susan Fenimore Cooper's *Rural Hours*." In *Susan Fenimore Cooper: New Essays on Rural Hours and Other Works*, edited by Rochelle Johnson and Daniel Patterson, 154–168. Athens, GA: University of Georgia Press, 2001.
Fowler, Robert Booth. *The Greening of Protestant Thought*. Chapel Hill: University of North Carolina Press, 1995.
Gatta, John. *Making Nature Sacred: Literature, Religion, and Environment in America from the Puritans to the Present*. Oxford: Oxford University Press, 2004.
Gianquitto, Tina. "The Noble Designs of Nature: G-d, Science, and the Picturesque in Susan Fenimore Cooper's *Rural Hours*." In *Susan Fenimore Cooper: New Essays on Rural Hours and Other Works*, edited by Rochelle Johnson and Daniel Patterson, 169–190. Athens, GA: University of Georgia Press, 2001.
Gottlieb, Roger S. *A Greener Faith: Religious Environmentalism and Our Planet's Future*. Oxford: Oxford University Press, 2006.
Herford, R. Travers. Edited and translated. *The Ethics of the Talmud: Sayings of the Fathers*. New York: Schocken Books, 1971.

Johnson, Rochelle and Daniel Patterson, eds. *Susan Fenimore Cooper: New Essays on Rural Hours and Other Works*. Athens, GA: University of Georgia Press, 2001.
Johnson, Rochelle. "Placing *Rural Hours*." In *Reading under the Sign of Nature: New Essays in Ecocriticism*, edited by John Tallmadge and Henry Harrington, 64–84. Salt Lake City, UT: University of Utah Press, 2000.
Lear, Linda. "Silent Spring." *The Life and Legacy of Rachel Carson*. rachelcarson. org/SilentSpring.aspx.
Leopold, Aldo. *A Sand County Almanac and Sketches Here and There*. New York: Oxford University Press, 1949.
Lovelock, James. "The Fallible Concept of Stewardship of the Earth." In *Environmental Stewardship: Critical Perspectives – Past and Present*, edited by R. J. Berry, 106–111. London: T&T Clark International, 2006.
Lovelock, James. *Gaia: A New Look at Life on Earth*. Oxford: Oxford University Press, 2000.
Maddox, Lucy B. "Susan Fenimore Cooper's Rustic Primer." In *Susan Fenimore Cooper: New Essays on Rural Hours and Other Works*, edited by Rochelle Johnson and Daniel Patterson, 83–95. Athens, GA: University of Georgia Press, 2001.
Magee, Richard M. "Sentimental Ecology: Susan Fenimore Cooper's *Rural Hours*." In *Such News of the Land: U.S. Women Nature Writers*, edited by Thomas S. Edwards and Elizabeth A. De Wolfe, 27–36. Hanover, NH: University Press of New England, 2001.
Northcott, Michael S. *The Environment and Christian Ethics*. Cambridge: Cambridge University Press, 1996.
Patterson, Daniel. Introduction. In Susan Fenimore Cooper, *Essays on Nature and Landscape*, edited by Rochelle Johnson and Daniel Patterson, xiii–xxxii. Athens, GA: University of Georgia Press, 2002.
Snyder, Gary. "Nothing Need Be Done," interview by Julia Martin. *Tricycle*. Spring 2015. https://tricycle.org/magazine/nothing-need-be-done. Adapted from Julia Martin and Gary Snyder. *Nobody Home: Writing, Buddhism, and Living in Places*. San Antonio, TX: Trinity University Press, 2014.
Snyder, Gary. *Turtle Island*. New York: New Directions, 1974.

Part II

Dystopian Visions of Past, Present, and Future

4 Monstrous Stewardship and the Plantation in Charles Chesnutt's "The Goophered Grapevine"

Matthew Wynn Sivils

The image of the outwardly bucolic slave plantation as the site of a horrifying amalgam of racial oppression and environmental exploitation emerges early in the American literary tradition. One of the most pronounced examples appears in "Letter IX" from J. Hector St. John de Crèvecoeur's *Letters from an American Farmer* (1782), in which Crèvecoeur's narrator, a northern farmer named James, visits slave-holding Charles Town, South Carolina only to conclude that in such places "though surrounded with the spontaneous riches of nature… we find the most wretched people in the world."[1] James describes witnessing "showers of sweat and of tears which from the bodies of Africans daily drop and moisten the ground they till,"[2] and he ends his commentary with a chilling portrait of how the chattel slave system makes the land itself party to the subjugation and destruction of the black person's body – all to satisfy the greed of the often-absent plantation owner. James writes of the irony of the situation, that "the extreme fertility of the ground always indicates the extreme misery of the inhabitants!"[3] Ultimately, Crèvecoeur imagines the plantation as a grotesque crossroads between a pastoral fantasy and the all-too-real nightmare of chattel slavery, one that quickly sheds its superficial charm to reveal a well-tended monstrosity.

Indeed, fueled by the cultivation of labor-intensive, soil-eroding crops such as cotton, chosen for their ability to justify the continuation of an enslaved workforce, the plantation might best be described as kind of monster, or at least as the embodiment of a monstrous system of human and environmental exploitation. Defining the monster as a cultural construct, Jeffrey Jerome Cohen writes, "The geography of the monster is an imperiling expanse, and therefore always a contested cultural space."[4] The plantation is just such a contested space – and like all monsters it cannot die. Reborn one generation to the next – from chattel slavery to post-bellum sharecropping – the plantation, like some Hollywood zombie, rises from the grave clad in whatever garb best suits the cultural moment.

In this essay I examine Charles W. Chesnutt's "The Goophered Grapevine" as a tale that imagines the plantation as the site of this hideously realized, continually resurrected form of cultivation. Paul Outka writes that slavery was "coextensive with white stewardship of a pastoral landscape,"[5] and Chesnutt portrays the plantation landscape as the ultimate product of this "white stewardship." His narrative peels back the Arcadian veil, exposing the sinister agro-ecological system that devours slaves and degrades the environment to enrich the white steward. When we read such nineteenth-century texts in association with recent ecogothic literary criticism and monster theory, a more accurate portrait of the plantation emerges: the plantation not as a romantic garden, but as an environmental grotesque born of a monstrous stewardship.

When Chesnutt's "The Goophered Grapevine" appeared in the *Atlantic Monthly* in 1887 few readers would have suspected it was anything more than another example of the regional dialect tale so popular at the time. Its setting, a romantically decaying plantation located in reconstruction-era North Carolina, checked the genre's usual boxes, and its main characters – a northern carpetbagger named John and an aged former slave named Uncle Julius with a knack for storytelling – are stock figures contemporary readers had met in one form or another in previous tales by writers such as Thomas Nelson Page and Joel Chandler Harris. So, Chesnutt's frame tale about a slave whose health is magically linked to an old plantation grapevine looked right at home in the pages of the 1887 *Atlantic Monthly*. But unknown to readers at that time, Chesnutt was, unlike Page and Harris, an African American, and "The Goophered Grapevine" conveys a far more racially nuanced stance than similar genre tales by those authors.

Regional, "local color," stories were then at the apex of their popularity, resonating with an American readership experiencing an intimidating amount of social change. This readership, almost entirely composed of a growing white middle class, watched as the nation continued to develop from a disconnected agrarian state into an increasingly de-localized industrial nation. A dizzying array of technologies and innovations (railroads, telegraph lines, telephones) began to draw together far flung and regionally distinct communities. "Paradoxically," writes Richard Brodhead,

> this process of delocalization helped produce a form of writing devoted to featuring local difference, so that the literature of local color emerged as the dominant American literary genre in the same decades as did the transcontinental railroad and Standard Oil.[6]

At the same time that the country shifted from a localized rural identity to a de-centralized urban one, another effort – one bent on dividing America across racial lines – began to make itself known.

The last decades of the nineteenth century marked the beginning a new, post-reconstruction-era push to not only degrade but to entirely disenfranchise African Americans through a multifaceted effort to terrorize them while robbing them of their legal protections. Pockmarked with separate schools, hospitals, cemeteries, bathrooms, water fountains, and parks, America became defined not by unity but by division, as states, especially those in the South, curtailed or forbade social interactions between blacks and whites. Along with these formalized measures was the ever-present threat of violence. One of the most horrifying examples of which was spectacle lynching, in which white mobs assembled to participate in the extra-judicial torture and execution of a black person – often for dubious or outright false accusations. As Richard M. Perloff writes,

> Typically, the victims were hung or burned to death by mobs of White vigilantes, frequently in front of thousands of spectators, many of whom would take pieces of the dead person's body as souvenirs to help remember the spectacular event.[7]

Between 1882 and 1968 over 3,445 African Americans were murdered by lynching, of which 1,645 were killed in the 1880s and 1890s – the decades that saw the publication of many of Chesnutt's short stories.[8] Works of fiction by writers such as Chesnutt who lived through this period of escalating racial violence prove especially instructive, particularly in terms of how Chesnutt employs an imagined plantation environment to comment upon the persistence of racial oppression in the otherwise increasingly modern nation.

"Southern Gothic," argues Maisha L. Wester, "can be understood as a genre that is aware of the impossibility of escaping racial haunting and the trauma of a culture that is not just informed by racial history, but also haunted and ruptured by it."[9] Writing within this southern gothic mode, Chesnutt's conjure tales (of which "The Goophered Grapevine" was the first) stand as racial ghost stories ripe with references to African American folk traditions. These stories of magical transformations and insidious curses often involve the metamorphosis of humans into plants or animals within the plantation setting. Rather than the supernatural, it is the unnatural practice of white oppression that forms the actual, unifying element of terror found in these narratives. African American stories of the supernatural, in the words of Tiya Miles,

> represent the past and serve as a marker of historical memory, but they represent a past with teeth, a past that is rife with trauma and the terror that characterizes the experience of an oppressed people… Their presence signals a need for those in the present to deal directly with a dangerous past that will not rest.[10]

In this way, Chesnutt leverages tropes common to the often-racist and past-obsessed regional fiction of his day to comment on the disturbing state of racial injustice in post-bellum America. One of the most powerful of these tropes is that of the gothic monster, an entity that represents in its variety of forms the repressed anxieties of our culture. Stacy Alaimo writes, "landscapes, unlike monsters, are devoid of agency; they neither chase or devour us,"[11] but Chesnutt's imagined plantations complicate this idea. Invoking the specter of racialized monsters, specifically those that arise from the horror of the southern plantation system, Chesnutt calls attention to how, despite its modern pretenses at the dawn of the twentieth century, the U.S. remained a nation mired in the trauma of racial injustice.

In crafting a tale that melds the memory of slavery with a portrait of present-day plantation life, he underscores the dangers of romanticizing the plantation while also highlighting ways the specter of racial injustice haunts the American mind. In studying the intersections between the concept of property and the horrors of slavery in southern gothic literature, Sarah Gilbreath Ford writes,

> Out of all of slavery's horrors, from violence and sexual assault to the separation of families, the status of enslaved people as property may seem academic and too far removed from the direct bodily harm of slavery to be the central catalyst for haunting. It is, however, the status of property that allows the perpetration of all of the other terrors.[12]

And María del Pilar Blanco writes that "rather than reading haunting and ghosts as past conundrums in search of closure," we should view "these phenomena in literature and film as experiments in a prolonged evocation of future anxieties."[13] Chesnutt's account of African American folkways within the plantation setting presents readers with just such an instance of cultural haunting. Drawn from a tableau of conceits common to nineteenth-century regional stories, Chesnutt's "The Goophered Grapevine" consists of a frame tale set on an antebellum plantation in North Carolina. It is predominantly narrated in African American dialect by a former slave named Uncle Julius who tells his story to a carpetbagger couple from Ohio (named John and Annie). Keen to buy the plantation and establish a vineyard there, the northerners soon learn that Julius was enslaved on the plantation and that he has continued to live there in the years following the war. Julius, who wishes to dissuade the pair from buying the plantation (ostensibly to preserve his own side business of selling the grapes that grow wild there), spins a yarn about a how the plantation grapevines suffer from a long-standing curse. He tells the couple that the plantation's owner, Mars Dugal' McAdoo, fed up with his slaves stealing his grapes, hired a local conjure woman named Aunt Peggy to put a "goopher" (a curse) on the grapevines so that any slave who ate from the vine would die within the span of a year.[14]

The local slaves – respecting Aunt Peggy's reputation as a conjurer – stop eating the grapes. However, a new slave named Henry, who is ignorant of the curse, eats some of the grapes. Taking pity on Henry, Aunt Peggy has him drink a concoction of her own making and instructs him to anoint his head with sap from the grapevine each spring. Henry follows her instructions, and not only is the curse abated, but as the grapevines flourish that spring so does Henry's health. With the coming of winter and the dormancy of the grapevine, Henry's vigor likewise diminishes, and the cycle repeats the following spring. Once aware of this unique situation, Mars Dugal' decides he can make a hefty profit by selling Henry in the spring (while at the peak of his magical vivacity), and then re-purchasing the weakened slave (at a much-reduced price) each winter. The scam works season after season, until one day a northern con man selling winepresses visits the plantation and convinces Mars Dugal' that he could produce more grapes if he adopted the northerner's method of tending the vines, which involves cutting them back, digging around the roots, and applying a mixture of various fertilizers. At the start of spring the grapevines grow well but they soon wither and die. Henry, whose fate is tied to that of the vines, also grows ill and dies. At the story's conclusion, an unfazed John buys the plantation and ends by bragging that under his stewardship it has become a successful business.

Written to an *Atlantic Monthly* audience that had no idea its author was an African American, it is compelling to read "The Goophered Grapevine" as a revisionist, even subversive, text draped in the disguise of the then-popular regional dialect tale. In this regard, Chesnutt's fiction prefigures a gradual twentieth-century shift in accounts of life under slavery. As one historian describes it, this shift is characterized by a reinterpretation of "the lives of enslaved people" in which "writers of revisionist studies celebrated an autonomous slave culture that flourished despite the horrors of bondage."[15] Chesnutt, for example, describes the slaves in far more humanized terms than in similar genre stories by his white contemporaries. His attention to the nuances of African American speech, often rendered poorly in dialect by white authors of the time, also contributes to the veracity of the characters. Additionally, Chesnutt subverts the typical romantic portrayal of the plantation itself when, in the postbellum frame tale, John sets this scene as he and his wife arrive at the plantation:

> We drove between a pair of *decayed* gateposts – the gate itself had long since disappeared – and up a straight sandy lane, between two lines of *rotting* rail fence, partly concealed by jimsonweeds and briers, to the open space where a dwelling-house had once stood, evidently a spacious mansion, if we might judge from the *ruined* chimneys that

were still standing, and the brick pillars on which the sill rested. The home itself, we had been informed, had fallen a *victim* to the fortunes of war.[16]

While a modicum of artful ruin is a common ingredient of romantic scenes, John's description of the plantation goes well beyond that threshold. Chesnutt, through John, paints a landscape cast almost entirely in shades of decay. The front gate is absent, and, standing out from the rotting vestiges of the plantation's former glory, the "ruined chimneys" serve as the last remnants of the otherwise destroyed mansion.

Ruin and absence define the man-made elements of Chesnutt's plantation, but the opposite condition best describes his portrayal of the vegetation that greets the two northerners. In John's scene-setting description, what has managed to thrive in the absence of stewardship is the titular grapevine itself:

> The estate had been for years involved in litigation between disputing heirs, during which period shiftless cultivation had well-nigh exhausted the soil. There had been a vineyard of some extent on the place, but it had not been attended to since the war, and had lapsed into utter neglect. The vines – here partly supported by decayed and broken-down trellises, there twining themselves among the branches of the slender saplings which had sprung up among them – grew in wild and unpruned luxuriance, and the few scattered grapes they bore were the undisputed prey of the first comer.[17]

Chesnutt makes smart use of his carpetbagger narrator to drive home a superficial view of the plantation as a place simply neglected by past owners and thus primed for future economic gain. When John looks at the plantation he sees dollar signs but misses those other signs of human and environmental exploitation that Uncle Julius can read all too clearly. In the absence of stewardship the vine has grown into a state of "wild and unpruned luxuriance," but Chesnutt's portrayal of the grapevine bears a more unsettling implication. As Cohen writes,

> The monster is... an embodiment of a certain cultural moment – of a time, a feeling, and a place... [It] inhabits the gap between the time of upheaval that created it and the moment into which it is received, to be born again.[18]

Like the grapevine, and despite the decay of the formal slave system, the oppression it embodies lives on in a different but no less robust form. Rooted in the soil of antebellum oppression, the vine curls around the "decayed... trellis" and the "sender saplings" alike, its tendrils joining the plantation's past with the present. An agricultural ghost haunting the plantation, the grapevine embodies the persistence of racial oppression that reaches its feelers from across the antebellum era into a post-bellum world plagued by predatory sharecropping, segregation, Jim Crow, and lynching. As Uncle Julius's tale reveals, this seemingly immortal grapevine is rooted in the horrors of a plantation past that keeps arising, keeps spreading, haunting future generations who would rather forget. In "The Goophered Grapevine," this recognition of the recurring monstrosity of the plantation system emerges most prominently at the juncture between its short-sighted agricultural stewardship and the horrifying cruelty associated with racial oppression. Chesnutt's gothic tale relates in fanciful terms the ludicrous but all-too-real horror of chattel slavery: that a person may become figuratively transformed into a non-human, a slave, and in so becoming is further transformed into a commodity to be bought, sold, and ultimately consumed.

Leveraging the allegorical power of the conjure tale, Chesnutt takes advantage of the genre's ability to give voice to otherwise muted implications of racial oppression. Describing the doomed Henry's seasonal transformation, Julius – in an African American dialect common to regional stories of the day – notes not only how the slave becomes magically linked to the grapevines but also how, through this curse, he becomes an even more valuable commodity.

> Henry 'n'int his head wid de sap out'n de big grapevime des ha'f way 'twix' de quarters en de big house... But the beatenes' thing you eber see happen ter Henry. Up ter dat time he wuz ez ball ez a sweeten' tater, but des ez soon ez de young leaves begun ter come out on de grapevimes, de ha'r begun to grow out on Henry's head, en by de middle er de summer he had de bigges' head er ha'r on de plantation... Henry's h'ar begun to quirl all up in little balls, des like dis yer reg'lar grapy h'ar, en by de time de grapes got ripe his head look des like a bunch er grapes... But dat wa'n't de quares' thing 'bout de goopher. When Henry come ter de plantation, he wuz gittin' a little ole and stiff in de j'ints. But dat summer he got... spry and libely... En nex' spring, w'en he rub de sap on ag'in, he got young ag'in, en so soopl en libely dat none er de young niggers on the plantation couldn' jump, ner dance, ner hoe ez much cotton ex Henry.[19]

Julius' description of Aunt Peggy's conjure remedy involves a series of transformations, each characterized by a dual process of hybridization and dehumanization. Henry, already dehumanized by his designation as a slave, is further transformed when he takes on a resemblance to the grapevine as well as its seasonal vitality. Stating that Henry arrived on the plantation "ez ball ez a sweeten' tater," Julius hints that even before the curse Henry was already something of a crop, an economic commodity bound to the land. "Every monster is... a double narrative," writes

Cohen, "two living stories: one that describes how the monster came to be and another, its testimony, detailing what cultural use the monster serves."[20] Henry, a monstrous victim of slavery, represents how dehumanization breeds further torments. Aunt Peggy's magical remedy makes him strong each spring, but it also makes him a more valuable commodity and all the more a victim of the predatory system. Henry's untimely death – the result of Mars Dugal'"s foolish stewardship – runs counter to bucolic plantation tales in which the white plantation owner serves as a beneficent steward of the land, as well as, in the racist calculus of the day, the slaves or African American sharecroppers whom they exploited for their labor. In fact, the portrayal of how a Yankee con-man's trick results in Mars Dugal' killing his own grapevine and, by association, the lucrative Henry, demonstrates how the larger economic system transcends geographical and ideological borders, making those ostensibly outside the system (i.e., the northern states) nevertheless complicit in its horrors.

Employing Julius' oft-ridiculous characters in service to this larger national allegory, the ending of Chesnutt's story directly comments on the northern influence upon the southern plantation. Noting Julius's tendency to derive material gain from the stories he tells, Sarah Gilbreath Ford writes,

> Where Julius's power truly lies is… not in the accumulation of things; he in fact depicts gain and greed as the motivation for much suffering in slavery. The power is in his ability to tell the story of the people he knew and the atrocities of the slave system.[21]

The northern influence on this system most emphatically enters the story in the form of the Yankee salesman who visits the plantation with a scheme to cheat Mars Dugal' by appealing to his considerable greed. In these passages Chesnutt highlights the unsavory similarity between these two men, characterizing the southern plantation owner (himself a con man in his dishonest re-selling of Henry) and the Yankee as each given to promoting the exploitation of people and the land for a quick buck. As Julius relates, the Yankee

> come down ter Norf C'lina fer ter l'arn de w'ite folks how to raise grapes en make wine. He promus Mars Dugal' he c'd make de grapevimes b'ar twice't ez many grapes, en dat de noo winepress he wuz a-sellin' would make mo' d'n twice't ez many gallons er wine.[22]

Chesnutt's commentary hinges upon the portrayal of a mutually beneficial system of exploitation: the southern plantation owner supplies the crops, and the northern market furnishes the demand. Additionally, the North (here in the form of the Yankee) engages in the ostensible sale of technology and expertise. That the conman's disastrous advice indirectly leads to Henry's death serves as a recognition of how the tendrils of the plantation intertwined with the nation's socio-economic system.

The Yankee's act of tricking the plantation owner into simultaneously destroying his prized grapevine and his magically remunerative slave is ultimately emblematic of the north's assault upon the southern economy in the aftermath of the Civil War. In fact, Chesnutt concludes Julius's story by comically linking the Yankee's trickery to the Civil War, stating that

> W'en de wah broke out, Mars Dugal' raise' a comp'ny, en went off ter fight de Yankees. He say he wuz mighty glad dat wah come, en he des want ter kill a Yankee fer eve'y dollar he los' 'long er dat grape-raisin' Yankee. En I 'spec' he would 'a' done it, too, ef de Yankees hadn' s'picioned sump'n, en killed him fus.'[23]

Following this humorous end to Julius' tale, John concludes by noting that despite the former slave's claim that the plantation's grapevines were cursed, he nevertheless purchased the plantation and his business has since flourished. Chesnutt thus effectively echoes, in John's present-day portion of the story, a similar situation to that described by Julius in his antebellum account. In both the antebellum and the post-bellum portions of Chesnutt's tale, a northern businessman disrupts the southern status quo through what is more or less an economically self-interested scheme, a situation in which Chesnutt ultimately points out not only the cyclical and predatory nature of white stewardship in the plantation system but also the way the north remained complicit in that system.

Chesnutt's story makes clear that racial oppression does not require any magical intervention to render a person inhuman; it merely needs the social conjuration of slavery, or, in the post-bellum world, the ever-present insult of Jim Crow laws backed by the threat of violence. It holds the power to strip away personhood, to render a human into a commodity. And as Chesnutt's tale suggests, emancipation brought about little change for those subjected to plantation labor. Ronald L. F. Davis notes,

> The life and work experiences of the black agricultural workers who labored as wage hands, as tenants, or as sharecroppers differed little regardless of the various terms that legally defined their status. The plantation still existed as a legal entity under the ownership or control of a planter landlord or merchant.[24]

The horror of tales like Chesnutt's resides in part in the nightmarish reality from which they spring and the recognition that little really changed in the decades following emancipation. Marked by violence and denigration, the ever-present specter of racism that haunts such narratives gives rise to a variety of particularly real, particularly relevant monsters. Slave and slave owner alike are transformed into monstrous figures straddling the chasm between the human and something else. Ultimately, the land of the plantation – the stage upon which this horrifying perversion of stewardship is enacted – becomes itself a grotesque victim, the diseased host for a parasitic system.

Notes

1. J. Hector St. John de Crèvecoeur, *Letters from an American Farmer* and *Sketches of Eighteenth-Century America*, edited by Albert E. Stone (New York: Penguin), 176.
2. Crèvecoeur, *Letters from an American Farmer*, 168.
3. Crèvecoeur, *Letters from an American Farmer*, 176.
4. Jeffrey Jerome Cohen, "Monster Culture (Seven Theses)," in *Monster Theory: Reading Culture*, ed. Jeffrey Jerome Cohen (Minneapolis, University of Minneapolis Press, 1996), 7.
5. Paul Outka, *Race and Nature: From Transcendentalism to the Harlem Renaissance* (New York: Palgrave, 2008), 103.
6. Richard H. Brodhead, "Introduction," in *The Conjure Woman and Other Conjure Tales*, ed. Richard H. Brodhead. (Durham, NC: Duke University Press, 1993).
7. Richard M. Perloff, "The Press and Lynchings of African Americans," *Journal of Black Studies* 30, no. 3 (January 2000), 315.
8. Perloff, "The Press and Lynchings of African Americans," 315.
9. Maisha L. Wester, *African American Gothic: Screams from Shadowed Places* (New York: Palgrave Macmillan, 2012), 25.
10. Tiya Miles, *Tales from the Haunted South: Dark Tourism and Memories of Slavery from the Civil War Era* (Chapel Hill: University of North Carolina Press, 2015), 126.
11. Stacy Alaimo, "Discomforting Creatures: Monstrous Nature in Recent Film," in *Beyond Nature Writing Expanding the Boundaries of Ecocriticism*, eds. Karla Armbruster and Kathleen R. Wallace (Charlottesville: University of Virginia Press, 2001), 283.
12. Sarah Gilbreath Ford, *Haunted Stewardship: Slavery and the Gothic* (Jackson: University Press of Mississippi, 2021), 6.
13. María del Pilar Blanco, *Ghost-Watching American Modernity: Haunting, Landscape, and the Hemispheric Imagination* (New York: Fordham University Press, 2012), 7.
14. Charles W. Chesnutt, "The Goophered Grapevine," in *The Conjure Woman and Other Conjure Tales*, eds. Richard H. Brodhead (Durham, NC: Duke University Press, 1993), 37.
15. Damian Alan Pargas, *The Quarters and the Fields: Slave Families in the Non-Cotton South* (Gainesville: University Press of Florida, 2010), 6.
16. Chesnutt, "The Goophered Grapevine," 33–34; italics added.
17. Chesnutt, "The Goophered Grapevine," 33.
18. Cohen, "Monster Culture (Seven Theses)," 4.
19. Chesnutt, "The Goophered Grapevine," 39.
20. Cohen, "Monster Culture (Seven Theses)," 13.
21. Sarah Ford Gilbreath, *Tracing Southern Storytelling in Black and White* (Tuscaloosa: University of Alabama Press, 2014), 26–27.
22. Chesnutt, "The Goophered Grapevine," 41.
23. Chesnutt, "The Goophered Grapevine," 42–43.

24 Ronald L. F. Davis, "The Plantation Lifeworld of the Old Natchez District: 1840–1880," in *Plantation Society and Race Relations: The Origins of Inequality*, eds. Thomas J. Durant Jr. and J. David Knottnerus (Westport, CT: Praeger, 1999), 177–178.

Bibliography

Alaimo, Stacy. "Discomforting Creatures: Monstrous Nature in Recent Film." In *Beyond Nature Writing Expanding the Boundaries of Ecocriticism*, edited by Karla Armbruster and Kathleen R. Wallace, 279–294. Charlottesville: University of Virginia Press, 2001.
Blanco, María del Pilar. *Ghost-Watching American Modernity: Haunting, Landscape, and the Hemispheric Imagination*. New York: Fordham University Press, 2012.
Brodhead, Richard H. "Introduction." In *The Conjure Woman and Other Conjure Tales*, edited by Richard H. Brodhead, 1–21. Durham, NC: Duke University Press, 1993.
Chesnutt, Charles W. "The Goophered Grapevine." In *The Conjure Woman and Other Conjure Tales*, edited by Richard H. Brodhead, 31–43. Durham, NC: Duke University Press, 1993.
Cohen, Jeffrey Jerome. "Monster Culture (Seven Theses)." In *Monster Theory: Reading Culture*, edited by Jeffrey Jerome Cohen, 3–25. Minneapolis, University of Minneapolis Press, 1996.
Crèvecoeur, J. Hector St. John de. *Letters from an American Farmer* and *Sketches of Eighteenth-Century America*, edited by Albert E. Stone. New York: Penguin, 1986.
Davis, Ronald L. F. "The Plantation Lifeworld of the Old Natchez District: 1840–1880." In *Plantation Society and Race Relations: The Origins of Inequality*, edited by Thomas J. Durant, Jr. and J. David Knottnerus, 165–179. Westport, CT: Praeger, 1999.
Ford, Sarah Gilbreath. *Haunted Stewardship: Slavery and the Gothic*. Jackson: University Press of Mississippi, 2021.
Ford, Sarah Gilbreath. *Tracing Southern Storytelling in Black and White*. Tuscaloosa: University of Alabama Press, 2014.
Miles, Tiya. *Tales from the Haunted South: Dark Tourism and Memories of Slavery from the Civil War Era*. Chapel Hill: University of North Carolina Press, 2015.
Outka, Paul. *Race and Nature: From Transcendentalism to the Harlem Renaissance*. New York: Palgrave, 2008.
Pargas, Damian Alan. *The Quarters and the Fields: Slave Families in the Non-Cotton South*. Gainesville: University Press of Florida, 2010.
Perloff, Richard M. "The Press and Lynchings of African Americans." *Journal of Black Studies* 30, no. 3 (January 2000): 315–330.
Wester, Maisha L. *African American Gothic: Screams from Shadowed Places*. New York: Palgrave Macmillan, 2012.

5 Human Stewardship and "Reproductive Futurism" in Dystopian Fiction

Pramod K. Nayar

The dystopian fiction that constitutes the subject of this chapter, *Never Let Me Go* (Kazuo Ishiguro, 2005), the *Xenogenesis* trilogy (Octavia E. Butler, 1987–1989) and *The Handmaid's Tale* (Margaret Atwood, 1986), invoke the full semantic scope of the term "steward," which means "guard," "overseer of workmen," "one who manages affairs of an estate on behalf of his employer" and a "person who supervises arrangements," all of which imply control, power and dominance.[1] I propose that stewardship in eugenic dystopias such as these, while directed at the future of the human race in terms of its very continuity and existence, foregrounds what Rebekah Sheldon following Lee Edelman terms "reproductive futurism." Sheldon has suggested that reproductive futurism "is a two-sided salvation narrative: someday the future will be redeemed of the mess our present actions foretell; until then, we must keep the messy future from coming by replicating the present through our children."[2]

Sheldon posits a link between economic and reproductive futurism in the form of a "somatic capitalism":

> somatic capitalism operates above and below the level of the individual subject to amplify or diminish specific bodily capacities. It siphons vitality rather than exerting discipline, swerves and harnesses existing tendencies rather than regulating their emergence. It differentially distributes exposures and zones of safety, but with the implicit acknowledgment that no system is ever really closed enough to be safe. Its accelerant is capital, and it rides on the profits to be reaped from catastrophe. It is an expression of the move from state biopolitics with its rhetoric of concern to neoliberal speculation. Its focus is on species as repositories of recombinant capacities.[3]

Sheldon's argument regarding the capitalization of life itself resonates with that of others, who trace the emergence of a bioeconomy around all biological processes and materials, from cells to blood, genes to surrogate wombs.[4] In such a bioeconomy, the maternal, the filial and the communal are embedded in the triad of state–corporate–medico-scientific structures, although this triad is mostly discernible by inference. We do not find clear distinctions between the Canadian, British or American social and political contexts in these texts, since they all propose totalitarian states (Gilead in Atwood, London of the future in Ishiguro, and the American continent in Butler).

However, the dystopian vision of Atwood, Butler and Ishiguro treats reproductive futurism as not (only) the creation of (improved) human bodies – in terms of cures for diseases, freedom from cancer, longevity, among others – but also, *contra* Sheldon, lacking a redemption from humanity's political and social pasts of violence, coercion, oppression and discrimination. In other words, the stewardship for "continuing" humanity is founded on structures of power and knowledge that remain scarred by the political and social pasts of humanity and does not ensure a redemptive future.

In Butler's *Xenogenesis* trilogy (*Dawn, Imago, Adulthood Rites*), set on an Earth two hundred and fifty years after a nuclear catastrophe, the Oankali, a race of beings from another planet, have arrived. They are "gene traders," who require human genes to continue their race. As they set about on a mission to mate with the survivors on Earth, they delete humanity's cancer-inducing genes, even as they absorb these genes as useful to their own continuity. Producing human–Oankali hybrid lifeforms ("constructs" as Butler terms them), the Oankali meet with some resistance from humans who do not wish such a miscegenated existence, and who eventually set up a colony on Mars. In Atwood's *The Handmaid's Tale*, the country of Gilead solves the problem of mass sterility by identifying "handmaids." Handmaids are allotted to specific politically and economically powerful families. On their days of fertility, the handmaid is raped (an event termed "Ceremony") by the "Commander," the head of the household, in the presence of their legal wives so that eventually the handmaid becomes pregnant. Once the child is born, he or she is integrated into the family, and the handmaid reallotted to another Commander. Ishiguro's *Never Let Me Go*, set in a future England, envisions an era when clones are created in large numbers as a matter of state policy. The clones are brought up in specialized schools until they reach a certain age when they start "donating": their vital organs are harvested to replace those damaged ones in "normal" humans. After their fourth such donation, the clones die ("complete"), until which time they are taken care of by "carers" who will themselves, one day, become "donors."

In dystopian fiction from the last two decades of the twentieth century and into the first decade of the twenty-first, knowledge, technology, genetic materials, and the regulation of all these, are directed at controlling a certain human future. There are, naturally, some differences. In Atwood, this vision is gender-centric, focusing on the reproductive enslavement of women for the supposed "needs" of the state. In Butler, while there is an emphasis on reproductive roles, this is more homogenized since *both* men and women of the human race are rendered into "mates" for the Oankali. In Ishiguro, the clones serve the human race. One could, of course, think in terms of a thematic of slavery running through these texts (as some critics have done): women as slaves in Atwood, the humans as reproductive vessels for Oankali in Butler (but in a symbiotic rather than a slave–master relation) and clones as slaves in Ishiguro. In this essay, however, my interest lies primarily in the common theme uniting the three: stewardship and the future of humanity.

They are all constitutive of stewardship systems, where the systems are not about preserving the earth or its non-human species – as the term often suggests – but about preserving the human occupants of the planet, and preserving them within a specific social structure. Stewardship entails control over and management of the human occupants of the planet, over the form of their "personhood" and the forms of kinship and social hierarchies they will be embedded in.

Precarious Heredity

Three forms of precarity that cut across generations of the human race and which call into question the future of the race may be identified in the novels under study: humans genetically prone to self-destruct as a race; reproductive crises; and mortality from diseases and/or environmental pollution. The documentation of a precarious heredity that endangers the future of humanity marks the novels' opening moves.

Humans, the alien Jdahya informs Lilith in *Dawn*, the first volume of Octavia Butler's *Xenogenesis* trilogy, have a "genetic inclination to grow [cancers]."[5] This is the human inheritance: of genetic material that endangers the individual, the lineage, and the race itself. Butler in *Dawn* elaborates the precarity that humans as a race carry within their *genes*, from the perspective of the Oankali (who can be male, female, or a third sex which is neither male nor female). Jhadya is speaking to Lilith, the human survivor who has been first "awakened":

> You have a mismatched pair of genetic characteristics. Either alone would have been useful, would have allowed the survival of your species. But the two together are lethal. It was only a matter of time before they destroyed you...
> You are intelligent... That's the newer of the two characteristics, and the one you might have put to work to save yourselves...
> You are hierarchical. That's the older and more entrenched characteristic. We saw it in your closest animal relatives and in your most distant ones. It's a terrestrial characteristic...

And a complex combination of genes that work together to make you intelligent as well as hierarchical will still handicap you whether you acknowledge it or not.[6]

There are several points of interest in the Oankali's description of the human race. First, of course, is the latent genetic feature – flaw – in the race, where the two genetic characteristics were bound to destroy the race because they occur and work in conjunction. If it were just one or the other, the race could have coped, but not when they occur together. Second, intelligence was a late arrival as a feature of the human race ("newer"), one which the humans had developed and put to work for its survival. Human intelligence, the Oankali propose, is epigenetic, and is a transmissible feature of the phenotype that is humanity. Intelligence is an add-on, acquired through training and therefore from the environment.[7] The human race, in Butler's vision, has inexplicably honed a feature – epigenetically speaking – that contradicts and conflicts with its genetic inheritance (which is, essentially, hierarchy).

For Atwood, the future of the human race is likewise precarious, and although written in the same decade but in different countries, the two authors demonstrate the same concern: human futures and survival. If Butler sees human survival as necessitating the intervention of an alien species, Atwood suggests that human survival will rely on a greater control over the women and the process of reproduction.

In Atwood's Gilead, the vast majority of the humans are sterile. It is also a singularly polluted world, and the toxins may have contributed to the sterility. Offred, through whom the events in Gilead are narrated, elaborates the features of this ecodystopian world:

> The air got too full, once, of chemicals, rays, radiation, the water swarmed with toxic molecules, all of that takes years to clean up, and meanwhile they creep into your body, camp out in your fatty cells. Who knows, your very flesh may be polluted, dirty as an oily beach, sure death to shore birds and unborn babies. Maybe a vulture would die of eating you... Women took medicines, pills, men sprayed trees, cows ate grass, all that souped-up piss flowed into the rivers. Not to mention the exploding atomic power plants, along the San Andreas fault, nobody's fault, during the earthquakes, and the mutant strain of syphilis no mold could touch.[8]

In the "Historical Notes" at the end of the novel we are given a set of probable causes for this widespread sterility of the humans. These "Notes" are supposedly the "partial transcript of the proceedings of the Twelfth Symposium on Gileadean Studies, held as part of the International Historical Association Convention, which took place at the University of

Denay, Nunavit, on June 25, 2195." The transcript is principally made up of a talk, "Problems of Authentication in Reference to *The Handmaid's Tale*" by "James Darcy Pieixoto, Director, Twentieth and Twenty-First Century Archives, Cambridge University, England." The Notes state:

> The reasons for this decline are not altogether clear to us. Some of the failure to reproduce can undoubtedly be traced to the widespread availability of birth control of various kinds, including abortion, in the immediate pre-Gilead period. Some infertility, then, was willed, which may account for the differing statistics among Caucasians and non-Caucasians; but the reasons for this decline are not altogether clear to us. Some of the failure to reproduce can undoubtedly be traced to the widespread availability of birth control of various kinds, including abortion, in the immediate pre-Gilead period. Need I remind you that this was the age of the R-strain syphilis and also the infamous AIDS epidemic, which, once they spread to the population at large, eliminated many young sexually active people from the reproductive pool? Stillbirths, miscarriages, and genetic deformities were widespread and on the increase, and this trend has been linked to the various nuclear-plant accidents, shutdowns, and incidents of sabotage that characterized the period, as well as to leakages from chemical and biological-warfare stockpiles and toxic-waste disposal sites, of which there were many thousands, both legal and illegal – in some instances these materials were simply dumped into the sewage system – and to the uncontrolled use of chemical insecticides, herbicides, and other sprays.[9]

In fact the condition is so pervasive that even mentioning the word ("sterile") is taboo. And this sterility is attributed to women. But the doctor whom Offred visits states the case more truthfully: "Most of those old guys can't make it any more," he says. "Or they're sterile." Offred is shocked at the candid admission:

> I almost gasp, he's said a forbidden word. Sterile. There is no such thing as a sterile man any more, not officially. There are only women who are fruitful and women who are barren, that's the law.[10]

The women who were fertile were therefore designated "handmaids," for the sole purpose of bearing children. Yet, notes Atwood's Offred recalling what the handmaids had been told by Aunt Lydia:

> Of course, some women believed there would be no future, they thought the world would explode. That was the excuse they used, says Aunt Lydia. They said there was no sense in breeding. Aunt Lydia's nostrils narrow: such wickedness. They were lazy women, she says.[11]

Those with specific defects are denied the right to procreate. Offred comments after observing Nick, one of the men serving the household of her allotted Commander: "low status: he hasn't been issued a woman, not even one. He doesn't rate: some defect."[12] One hears echoes of eugenics in that sentence, although Atwood presents it as a violation of reproductive rights.[13] There is a clear racial angle to the reproductive crisis as well, as the "Historical Notes" inform us of the situation in Gilead: "in an age of plummeting Caucasian birth rates, a phenomenon observable not only in Gilead but in most northern Caucasian societies of the time."[14]

Atwood foregrounds a universal sterility for humans, which would eventually wipe out the race unless radical measures, such as those in Gilead, are undertaken. Not unrelatedly, Ishiguro, two decades later, setting his novel in a London of the future, also envisions a human race depleted by its own pathologies and eventual mortality, and the precarious human race must be protected from their imminent deterioration and eventual mortality. As Miss Emily, one of the teachers at Hailsham, explains to the clones: "Their [the humans'] overwhelming concern was that their own children, their spouses, their parents, their friends, did not die from cancer, motor neurone disease, heart disease."[15]

The clones were to be the solution to the problem of the human race. When the people imagined a future, says Miss Emily, they "saw a new world coming rapidly. More scientific, efficient, yes. More cures for the old sicknesses."[16] It is in these contexts where human heredity and futures are at risk that mechanisms have been developed to ensure reproducibility, freedom from disease, and longevity. Ensuring that the human future has to proceed along lines determined by humans is a form of stewardship. The stewardship is directed at human futures, and ensuring that a future will exist for the human race through the management of genetic materials, in the main. For Atwood, Butler, and Ishiguro, stewardship is in effect a state-and-corporate investment in reproductive technologies.

The Stewardship of Heredity

You controlled both animals and people by controlling their reproduction.[17]

In the contexts of mortality, sterility, and pathologized human lives, the dystopian novels from Atwood, Butler, and Ishiguro foresee a stewardship of heredity itself through the management of reproduction and heredity.

Butler's dystopian vision includes two interrelated aspects of stewardship.

One, the Oankali ensure that the human control and stewardship over their genetic futures through their rudimentary (human) genetic engineering is erased. Whether as a consequence hierarchical thinking among the humans will persist or disappear is a moot point, since Butler's novels do not extend into the later generations of Oankali–human hybrids (or "Constructs," as Butler calls them). In other words, whether the human penchant for hierarchical thinking, a part of their genetic make-up, has been deleted through the Oankali's modifications of human genes is open to speculation. What we do know, however, is that the humans' control over their genetic futures is no longer a matter of right.

Second, through the hybridization and symbiogenesis, the human race is forced to move in directions that it may or may not want, but certainly in directions dictated by Oankali genes. Jdahya tells Lilith in *Dawn* that "correcting genes" have been introduced into her body so that she would not get cancer, even by accident.[18] Now, while this may be a good thing, the humans no longer possess an agency over their technologically determined futures. It is the Oankali genetic material that determines how human life on Earth will be lived hereafter. To word it differently, the stewardship of human genetic futures is now held by the Oankali genetic material inside the human–Oankali hybrids.

Intelligence and hierarchical thinking were in conflict because they contributed to the policing of difference, of social distinctions, and inequality among the humans. When the Oankali arrive and miscegenate with the humans, what is at stake, from the human perspective, is the loss of their (human) species sovereignty and, by extension, their social hierarchies. That is, the "species cosmopolitanism" engendered by the Oankali's symbiogenetic relations with humans calls into question the established hierarchies within the human race by subordinating the humans themselves to an alien lifeform.[19] The resisters (who eventually form a colony on Mars) who do not wish to mate with the Oankali are the stewards of the old humanity, "holdouts who cannot get over the alienness of the Oankali and reject their offer of [gene] trade in a futile effort to maintain the purity of the human race."[20]

If Butler treats human continuity in the face of the catastrophic state of the earth as made possible through alien intervention which would facilitate not only the deletion of cancer from human genetic material but also reproduction itself, Atwood's vision of human continuity centers on the role women will be forced to play toward this end. In Atwood's *Handmaid's Tale* the women are merely machines for reproduction, for heterosexual families. Aunt Lydia is educating the handmaids on the past, on human reproduction in history, and Offred recalls what was told to the girls:

> It used to be different… a pregnant woman, wired up to a machine, electrodes coming out of her every which way so that she looked like a broken robot, an intravenous drip feeding into her arm. Some man with a searchlight looking up between her legs, where she's been shaved, a mere beardless girl, a trayful of bright sterilized knives, everyone with masks on. A cooperative patient. Once they drugged women, induced labor, cut them open, sewed them up. No more. No anaesthetics, even. Aunt Elizabeth said it was better for the baby, but

also: *I will greatly multiply thy sorrow and thy conception; in sorrow thou shalt bring forth children.*[21]

Aunt Lydia's attempt to demonstrate how terrible it was for women in the past is part of the initiation of the handmaids, so that they accept their current reproductive slavery. In short, there are no options for fertile women in Atwood's vision, as Amin Malak argues in an essay on Atwood:

> The dire alternative for the handmaid is banishment to the Colonies, where women clean up radioactive waste as slave labourers. The dictates of state policy in Gilead thus relegate sex to a saleable commodity exchanged for mere minimal survival.[22]

In the future, in Atwood's vision, the fate of humankind is controlled and determined through a control over women's bodies. More importantly, Atwood emphasizes that the woman's body serves as a currency, or a commodity: "What we prayed for was emptiness, so we would be worthy to be filled: with grace, with love, with self-denial, semen and babies."[23] "So now that we don't have different clothes," says Offred to the Commander, "you merely have different women."[24] Their body is used by the handmaids themselves to pay their debts, pay their way to survival – which they do though giving birth. As Offred puts it, "It's up to me to repay the team, justify my food and keep."[25] Offred is also signalling the moral economy of Gilead, one that makes it imperative that she "repay" her team of handmaids by allowing herself to be a procreative subject. Offred describes her relationship with her master, the Commander:

> I don't love the Commander or anything like it, but he's of interest to me, he occupies space, he is more than a shadow. And I for him. To him I'm no longer merely a usable body. To him I'm not just a boat with no cargo, a chalice with no wine in it, an oven – to be crude – minus the bun. To him I am not merely empty.[26]

The capitalist economy of Gilead is the environment of the handmaids' reproductive function. That is, the science–capitalism link in Gilead's economy structures the life, reproductive functions and deaths of the handmaids. The state is the steward, which organizes the sexual and affective behaviour of the handmaids.

Aunt Lydia makes this point about the intergenerational and future-directed nature of state stewardship of women's bodies and reproductive functions when she says:

> You are a transitional generation... It is the hardest for you. We know the sacrifices you are being expected to make. It is hard when men revile you. For the ones who come after you, it will be easier. They will accept their duties with willing hearts.[27]

What Aunt Lydia is suggesting is that over generations, the women will adapt to the necessities of the new social order, and their own role within that order. As Offred declares, "I am a national resource," implying her role as a commodity and currency that enables the survival of the nation state itself. As Atwood puts it elsewhere in the novel in Offred's voice:

> A woman that pregnant doesn't have to go out, doesn't have to go shopping. The daily walk is no longer prescribed, to keep her abdominal muscles in working order. She needs only the floor exercises, the breathing drill. She could stay at her house. And it's dangerous for her to be out, there must be a Guardian standing outside the door, waiting for her. Now that she's the carrier of life, she is closer to death, and needs special security. Jealousy could get her, it's happened before. All children are wanted now, but not by everyone.[28]

Atwood shows how the fertile women are cared for: they are valuable *resources* for Gilead and must be guarded, yet have no other value or agency.

Similar care is exercised over the clones in Ishiguro. At Hailsham they are subject to health check-ups and provided the right nutrition and safety because they are necessary for the future of the human race. But despite this care, the clones are mere living cadavers who will one day donate their organs to the "normals": humans. The clones are expendable "persons" (in Ishiguro's post-human vision, the humanness of the clones is what is essentially tragic, as Karl Shaddox observed).[29] Like women in Gilead and humans to Oankali, the clones are resources to be exploited for somebody's safe future.

In Butler, the humans can continue to live on a devastated Earth if they co-reproduce with the Oankali. In Atwood, the handmaids are the possible bearers of human progeny in the future. In Ishiguro, while clone reproduction on behalf of the humans is not envisaged, human life *qua* life is indebted to the clones. All three represent humanity's survival as contingent upon a careful stewardship of reproduction: by aliens or by the humans themselves.

Stewardship, the Coercive Placental Economy, and Planetary Futures

> As in those pictures, those museums, those model towns, there are no children.[30]

The Oankali in Butler's trilogy delete those genetic markers that produce cancer and therefore accelerated mortality in humans. This re-geneticization as one can think of it, says one Oankali to Lilith, ensures "that you'll have a chance to live on your Earth – not just to die on it."[31] That is, humanity's future depends on acts of genetic engineering that will then inform the phenotype of future generations.

In the dystopian vision of Butler, Atwood, and Ishiguro, the connections between (i) humans and alien lifeforms, (ii) men, women and surrogate mothers, and (iii) humans and "their" clones respectively, are instances of a *coercive* placental economy, a concept adapted from Laurel Bollinger's reading of Octavia Butler. Bollinger argues that the "placental economy, like the placenta itself, offers a metaphor for exchanges in which both figures are protected from destructive fusion, while still fundamentally connected to one another," thereby implying both separateness and connectedness.[32] While Bollinger's emphasis is on Butler's interest in intersubjectivity and connectedness as a means of (human) survival, there are various disquieting aspects of this "connectedness" of the placental economy that demand attention.

Butler's reproductive futurism envisions human stewardship over species-heredity as controlled by the alien lifeform, where human–Oankali "interdependence and accessibility" is paid for with the "loss of historical identity," as Megan Obourn presciently describes it.[33] Donna Haraway had pointed to the problem at the heart of the trilogy: Lilith's pregnancy *for* the future of the human race was coerced, and *not* of her choice.[34] Critics such as Amanda Boulter, Michelle Green, and others have noted the rewriting of the slavery narrative in *Xenogenesis* with the humans serving as the animal-Other to the Oankali, just as Rachel Carroll describes the lives of the clones in *Never Let Me Go* as "biotechnological slavery."[35]

In all three novelists, the "persons" subject to the coercive placental economy, and who are seen as embodying the future of the human race, constantly seek the affirmation of their *locus standi* as persons, as maternal bodies, and family members in the new "versions" of kinship, whether between humans and alien sexual-partners (Butler), between handmaids and their male "owners" (Atwood) or between humans and clones (Ishiguro). Reminiscent of the quest for subjectivities and identities by minorities and marginalized in human history, the clones, the surrogate mothers, and genetically engineered humans in dystopian fiction seek validation and continuities. In the process, they discover that their historical identities, with class/race/gender differences open to exploitation and violence have been retained, if reconfigured for a "greater" cause. The coercive placental economy produces connectedness in which the connection and mutuality retain forms of subordination and inequalities, discovered when the "disposable" persons' search for affirmation, origins or kinship. To phrase it differently, stewardship directed at the future reproduces the violence immanent to the human race even as it (stewardship) hopes to alter the race's longevity, health and living conditions.

Three instances from the three texts respectively under discussion underscore the common theme of violence in stewardship of human futures.

First, sexuality and intimacy – defining features of human subjectivity – are all controlled. In *Never Let Me* Go, the cloned children are told at their school, Hailsham:

> Your lives are set out for you. You'll become adults, then, before you're old, before you're even middle-aged, you'll start to donate your vital organs. That's what each of you was created to do. You're not like the actors you watch on your videos, you're not even like me. You were brought into this world for a purpose, and your futures, all of them, have been decided.[36]

But adulthood includes sexuality and sexual rights, which the clones are denied. The teachers inform the students at Hailsham, reported to us through Kathy H (through whom we hear the story of the clones in the novel):

> Then suddenly... [Miss Emily] began telling us how we had to be careful about who we had sex with. Not just because of the diseases but because, she said, "sex affects emotions in ways you'd never expect." We had to be extremely careful about having sex in the outside world, especially with people who weren't students, because out there sex meant all sorts of things. Out there people were even fighting and killing each other over who had sex with whom. And the reason it meant so much – so much more than say, dancing or table tennis – was because the people out there were different from us students; they could have babies from sex.[37]

The clones are curious about their (proscribed) sexuality. Says Kathy H: "I also spent a lot of time re-reading passages from books where people had sex, going over the lines again and again, trying to tease out clues."[38] Such a management of sexuality and reproductive sexuality is aligned in Ishiguro's dystopian vision with the management of kinship. The clones' interest in questions of their sexual rights and anxieties over potential intimate relations is linked, as Rebecca Suter observes, to their larger questions of "unlived lives," and their origins from human "possibles."[39] That is, the interest in romance and sexuality among the clones is inseparable from their curiosity about their origins in human sexuality, romance, and structures of coupledom. Rachel Carroll argues in this connection: "sexuality becomes less an expression of desire, attachment or pleasure than another social discourse which must be learnt and emulated for the purposes of integration."[40] Thus, sexuality will have no affective investment possible in the future, in Carroll's reading of Ishiguro's vision:

> coupledom is understood less as an elective expression of a romantic or sexual affinity than as a necessary assumption of a culturally

coded set of practices: that is, as an index of successful assimilation into the world of the 'normals.'[41]

Kathy H enunciates their relation to and difference from their human originals along these lines in Ishiguro's novel:

> We certainly knew – though not in any deep sense – that we were different from our guardians, and also from the normal people outside; we perhaps even knew that a long way down the line there were donations waiting for us. But we didn't really know what that meant.[42]

Rosemary Rizq has suggested that Ishiguro "is defining our humanity precisely in terms of our kinship with clones – by suggesting that we are all copies of one sort or another (or copies of copies), because there was never anything original there in the first place."[43] While this implies the troubled connectedness – the clones are and are not connected to their human "possibles" who are, of course, untraceable – in the coercive placental economy of the future.

Second (to shift to Atwood's novel), we see how the existing structures of a heteronormative family and its continuity are preserved through the subjugation and sexual slavery of the handmaids. The handmaids have no possible role after giving birth. In each of these cases, we realize that the definitions of humanity and what we have inherited have altered irrevocably. There is a promise of some limited freedom from the chattel-slavery, and some agency, to the handmaids. Offred recalls what Aunt Lydia has promised the handmaids:

> The women will live in harmony together, all in one family; you will be like daughters to them, and when the population level is up to scratch again we'll no longer have to transfer you from one house to another because there will be enough to go round. There can be bonds of real affection, she said, blinking at us ingratiatingly, under such conditions. Women united for a common end! Helping one another in their daily chores as they walk the path of life together, each performing her appointed task. Why expect one woman to carry out all the functions necessary to the serene running of a household? It isn't reasonable or humane. Your daughters will have greater freedom. We are working towards the goal of a little garden for each one, each one of you.[44]

But handmaids such as Offred realise how dispensable they are once they fulfil their functions for the state. Offred thinks:

> And there will be family albums, too, with all the children in them; no Handmaids though. From the point of view of future history, this

kind, we'll be invisible. But the children will be in them all right, something for the Wives to look at, downstairs, nibbling at the buffet and waiting for the birth.[45]

If in Atwood the woman's sexuality has been subsumed to the needs of the state's men, in Butler, the sexuality of both men and women are controlled by the Oankali.

Third, in Butler's trilogy, the entire structure of extended families and kinship in Oankali–human families, which ensures the continuity of both species, are premised on the reluctant reproduction by the (human) women. In *Xenogenesis*, by the time we reach *Imago*, it is no longer possible to disentangle the human from their nonhuman origins. Dichaan explains to Akin that the humans had tried practicing isolation and exclusion to preserve their racial and ethnic "purity":

> The differences you perceive between Humans – between groups of Humans – are the result of isolation and inbreeding, mutation, and adaptation to different Earth environments," he said, illustrating each concept with quick multiple images. "Joseph and Lilith were born in very different parts of this world – born to long separated people."[46]

Lilith is given a view of the new family structures with the symbiogenesis of humans and Oankali: "Families will change, Lilith – are changing. A complete construct family will be a female, an ooloi, and children. Males will come and go as they wish and as they find welcome."[47]

In *Imago*, Butler outlines how new kinship structures based on mutuality could be built between and across different lifeforms. In an extended passage we are told how Lilith's cancer-genes were in fact instrumental in her aiding the Oankali, Nikanj, to live:

> Every child in the family had heard that story. One of Nikanj's sensory arms had been all but severed from its body, but Lilith allowed it to link into her body and activate certain of her highly specialized genes. It used what it learned from these to encourage its own cells to grow and reattach the complex structures of the arm. It could not have done this without the triggering effect of Lilith's genetic help.
> Lilith's ability had run in her family, although neither she nor her ancestors had been able to control it. It had either lain dormant in them or come to life in insane, haphazard fashion and caused the growth of useless new tissue. New tissue gone obscenely wrong.
> Humans called this condition cancer. To them, it was a hated disease. To the Oankali, it was treasure. It was beauty beyond Human comprehension.[48]

The configuration of the human family changes entirely, as does kinship, and the humans have little agency or choice in either domains.[49]

The genetic materials of the humans help the alien race in *Xenogenesis*. As surrogate (defined by Ruth McElroy as "one body standing in for another"[50]) *bodies*, the clones, the humans, and the handmaids displace that of the humans and human modes of reproduction and nurture (this especially in Butler), meanwhile also engendering new forms of familial and kinship relations.

In each case, the reconfigured sexual, familial, kinship, and communal relations in the coercive placental economy, while demonstrating dependence hinges on the (continuing) denial of sexual, reproductive, and affective agency. My reading is clearly more in line with Boulter, Obourn, and others who have detected, in the case of Octavia Butler, this violent rewriting of existing hierarchic and oppressive structures of class, race, and patriarchy. Obourn writes:

> As a woman whose role as mother is coerced and who identifies with a genealogy of other such women, Lilith cannot fully perform reproductive futurity. Additionally, Butler's model of the futuristic reproductive family unit presents desire and drive as part of coerced sexuality and motherhood and as part of the dynamic of car.[51]

With stewardship systems, such as those seen in these recent dystopian texts, historical identities of the maternal, the sexually active adult, the child, the nursemaid are reconfigured within structures of domination, slavery, and coercive reproduction/miscegenation.

Dystopian texts such as the ones discussed here do not envision planetary and human features, especially when humanity is managed, organized, and controlled in an attempt to ensure longevity, delayed mortality, and continuity, as being able to delete from its practices, even from memories, its past with all the violence, subjugation, and discriminatory binaries and Othering.

While the stewardship of the planet is integral to the possibility of human survival and continuity, Butler, Atwood, and Ishiguro remain focused on human-centered stewardship. To take the later novel first, Ishiguro demonstrates how human techno-scientific advancement would be directed at improving the continuity of human life and delaying death, albeit that this would entail the creation of a whole new lifeform: the clones. Cloning, which produces in the novel a lifeform designed to serve the human race, is therefore not just a technology of stewardship: it is also a technology of enslavement. In Atwood, the state decrees the instrumental use of women's fertility and bodies for the perpetuation of humanity. State control of fertile women, in Atwood's vision, is an extension of the gendered nature of human oppression, as historically practiced. For

Butler, the violent history of humanity that has reduced the earth to this devastated condition, can only minimally mitigated for human survival if aliens intervene – at some cost to the humans, of course.

Stewardship in these texts is about a possible human future, but the dystopian vision of this human stewardship, focused on the management of reproduction, underscores the inescapability of historical identities and practices. Human stewardship in Octavia Butler, Margaret Atwood, and Kazuo Ishiguro is thus envisioned as a violent, hierarchic, and exploitative prospect.

Notes

1 Margaret Atwood, *The Handmaid's Tale* (Toronto: Seal, 1986); Kazuo Ishiguro, *Never Let Me Go* (London: Faber and Faber, 2005), Octavia Butler, *Xenogenesis: Dawn* (New York: Warner, 1987), Octavia Butler, *Adulthood Rites* (New York: Warner, 1988) and Octavia Butler *Imago* (New York: Warner, 1989).
2 Rebekah Sheldon, *The Child to Come: Life after the Human Catastrophe* (Minneapolis: University of Minnesota Press, 2016), 35.
3 Sheldon, *The Child to Come*, 118.
4 See, among others, Kaushik Sunder Rajan, *Biocapital: The Constitution of Postgenomic Life* (Durham NC: Duke University Press, 2006); Pramod K. Nayar, "Precarious Lives in the Age of Biocapitalism," in *The Bloomsbury Handbook of Posthumanism*, ed. Mads Rosendahl Thomsen and Jacob Wamberg (London: Bloomsbury 2020).
5 Butler, *Dawn*, 29.
6 Butler, *Dawn*, 36–8.
7 Eva Jablonka says about this kind of inheritance: "[in] epigenetic inheritance the original environmental conditions *need not* be repeated because internal changes induced in the organism's physiology obviate the need for induction by the external stimulus (the external stimulus has been replaced by a persistent internal state)." Eva Jablonka, "Cultural Epigenetics," *The Sociological Review Monographs* 64, no. 1 (2016): 46, emphasis in original.
8 Atwood, *Handmaid's Tale*, 106.
9 Atwood, *Handmaid's Tale*, 286.
10 Atwood, *Handmaid's Tale*, 57.
11 Atwood, *Handmaid's Tale*, 286.
12 Atwood, *Handmaid's Tale*, 17.
13 Thomas Horan, *Desire and Empathy in Twentieth-Century Dystopian Fiction* (New York: Palgrave Macmillan, 2018), 179.
14 Atwood, *Handmaid's Tale*, 286.
15 Ishiguro, *Never Let Me Go*, 258.
16 Ishiguro, *Never Let Me Go*, 257.
17 Butler, *Adulthood Rites*, 203.
18 Butler, *Dawn*, 30.
19 Pramod K. Nayar, *Posthumanism* (Cambridge: Polity, 2012).
20 Jeffrey A. Tucker, "'The Human Contradiction': Identity and/as Essence in Octavia Butler's *Xenogenesis* Trilogy," *Science Fiction Studies* 37, no. 2 (2007): 167. Butler shows how humans cannot overcome their fascination with policing difference: even after the catastrophe resister villages are organized around racial and ethnic lines: Igbo, Hindu, Chinese and Spanish villages (*Adulthood Rites*).

21 Atwood, *Handmaid's Tale*, 108, emphasis in original.
22 Amin Malak, "Margaret Atwood's *The Handmaid's Tale* and the Dystopian Tradition," *Canadian Literature* 112 (1987): 9–16, 9.
23 Atwood, *Handmaid's Tale*, 182.
24 Atwood, *Handmaid's Tale*, 222.
25 Atwood, *Handmaid's Tale*, 127. For an analysis of the economics of the handmaids' lives, see Linda Myrsiades, "Law, Medicine, and the Sex Slave in Margaret Atwood's *"The Handmaid's Tale,"* *Counterpoints* 121 (1999): 219–245.
26 Atwood, *Handmaid's Tale*, 153.
27 Atwood, *Handmaid's Tale*, 111.
28 Atwood, *Handmaid's Tale*, 25–6.
29 Karl Shaddox, "Generic Considerations in Ishiguro's *Never Let Me Go*," *Human Rights Quarterly* 35, no. 2 (2013): 448–469.
30 Atwood, *Handmaid's Tale*, 23.
31 Butler, *Dawn*, 31.
32 Laurel Bollinger, "Placental Economy: Octavia Butler, Luce Irigaray, and Speculative Subjectivity," *Literature Interpretation Theory* 18 (2007): 330.
33 Megan Obourn, "Octavia Butler's Disabled Futures," *Contemporary Literature* 54, no.1 (2013): 112.
34 Donna Haraway writes: "it is a fatal pleasure that marks Lilith for the other awakened humans, even though she has not yet consented to." Donna Haraway, *Simians, Cyborgs, and Women: The Reinvention of Nature* (New York and London: Routledge, 1991), 229.
35 Rachel Carroll, "Imitations of life: cloning, heterosexuality and the human in Kazuo Ishiguro's *Never Let Me Go*," *Journal of Gender Studies* 19, no. 1 (2010): 62. See Michelle Erica Green, "'There Goes the Neighborhood': Octavia Butler's Demand for Diversity in Utopia," in *Utopia and Science Fiction by Women: Worlds of Difference*, ed. Jane L. Donawerth and Carol A. Kolmerten (Syracuse, NY: Syracuse University Press, 1994), 166–189; Amanda Boulter, "Polymorphous Futures: Octavia Butler's *Xenogenesis* Trilogy," in *American Bodies: Cultural Histories of the Physique*, ed. Tim Armstrong (New York, NY: New York University Press, 1996), 170–185, among others.
36 Ishiguro, *Never Let Me Go*, 80.
37 Ishiguro, *Never Let Me Go*, 82.
38 Ishiguro, *Never Let Me Go*, 97.
39 Rebecca Suter, "Untold and Unlived Lives in Kazuo Ishiguro's *Never Let Me Go*: A Response to Burkhard Niederhof," *Connotations* 21, nos. 2–3 (2011–2012): 391–406.
40 Carroll, "Imitations of Life," 66.
41 Carroll, "Imitations of Life," 67.
42 Ishiguro, *Never Let Me Go*, 69.
43 Rosemary Rizq, "Copying, Cloning and Creativity: Reading Kazuo Ishiguro's *Never Let Me Go*," *British Journal of Psychotherapy* 30, no.4 (2014): 530.
44 Atwood, *Handmaid's Tale*, 152.
45 Atwood, *Handmaid's Tale*, 214.
46 Butler, *Adulthood Rites*, 13.
47 Butler, *Adulthood Rites*, 11.
48 Butler, *Imago*, 30.
49 Theodora Goss and John Paul Riquelme describe the domestic set-up as follows: "The Oankali–human family structure and domestic arrangements are as follows. When a human couple of different sexes mates with the Oankali the enlarged group becomes internally differentiated and internally

double. There are two females and two males, as well as two pairs of partners of different sexes, one Oankali pair and one human pair. Every member of the pentadic family except the ooloi has a same-species partner of a different sex and a same-sex partner of a different species." They also note that this family set-up is based not on human preferences, but on Oankali triadic structures. Theodora Goss and John Paul Riquelme, "From Superhuman to Posthuman: The Gothic Technological Imaginary in Mary Shelley's *Frankenstein* and Octavia Butler's *Xenogenesis*," *Modern Fiction Studies* 53, no. 3 (2007): 447.
50 Ruth McElroy, "Whose Body, Whose Nation? Surrogate Motherhood and its Representation," *European Journal of Cultural Studies* 5, no. 3 (2002): 338.
51 Obourn, "Octavia Butler's Disabled Futures," 134.

Bibliography

Atwood, Margaret. *The Handmaid's Tale*. Toronto: Seal, 1986.
Bollinger, Laurel. "Placental Economy: Octavia Butler, Luce Irigaray, and Speculative Subjectivity." *Literature Interpretation Theory* 18 (2007): 325–352.
Boulter, Amanda. "Polymorphous Futures: Octavia Butler's *Xenogenesis* Trilogy." In *American Bodies: Cultural Histories of the Physique*, edited by Tim Armstrong, 170–185. New York, NY: New York University Press, 1996.
Butler, Octavia. *Adulthood Rites*. New York: Warner, 1988.
Butler, Octavia. *Imago*. New York: Warner, 1989.
Butler, Octavia. *Xenogenesis: Dawn*. New York: Warner, 1987.
Carroll, Rachel. "Imitations of life: cloning, heterosexuality and the human in Kazuo Ishiguro's *Never Let Me Go*." *Journal of Gender Studies* 19, no. 1 (2010): 59–71.
Goss, Theodora and John Paul Riquelme. "From Superhuman to Posthuman: The Gothic Technological Imaginary in Mary Shelley's *Frankenstein* and Octavia Butler's *Xenogenesis*." *Modern Fiction Studies* 53, no. 3 (2007): 434–459.
Green, Michelle Erica. "'There Goes the Neighborhood': Octavia Butler's Demand for Diversity in Utopia." In *Utopia and Science Fiction by Women: Worlds of Difference*, edited by Jane L. Donawerth and Carol A Kolmerten, 166–189. Syracuse, NY: Syracuse University Press, 1994.
Haraway, Donna. *Simians, Cyborgs, and Women: The Reinvention of Nature*. New York and London: Routledge, 1991.
Horan, Thomas. *Desire and Empathy in Twentieth-Century Dystopian Fiction*. New York: Palgrave Macmillan, 2018.
Ishiguro, Kazuo. *Never Let Me Go*. London: Faber and Faber, 2005.
Jablonka, Eva. "Cultural Epigenetics." *The Sociological Review Monographs* 64, no. 1 (2016): 42–60.
Malak, Amin. "Margaret Atwood's *The Handmaid's Tale* and the Dystopian Tradition." *Canadian Literature* 112 (1987): 9–16.
McElroy, Ruth. "Whose body, Whose nation? Surrogate Motherhood and its Representation." *European Journal of Cultural Studies* 5, no. 3 (2002): 325–342.
Myrsiades, Linda. "Law, Medicine, and the Sex Slave in Margaret Atwood's "*The Handmaid's Tale*"." *Counterpoints* 121 (1999): 219–245.

Nayar, Pramod K. "Precarious Lives in the Age of Biocapitalism." In *The Bloomsbury Handbook of Posthumanism*, edited by Mads Rosendahl Thomsen and Jacob Wamberg, 425–435. London: Bloomsbury 2020.
Nayar, Pramod K. *Posthumanism*. Cambridge: Polity, 2012.
Obourn, Megan. "Octavia Butler's Disabled Futures." *Contemporary Literature* 54, no. 1 (2013): 109–138.
Rajan, Kaushik Sunder. *Biocapital: The Constitution of Postgenomic Life*. Durham, NC: Duke University Press, 2006.
Rizq, Rosemary. "Copying, Cloning and Creativity: Reading Kazuo Ishiguro's *Never Let Me Go*." *British Journal of Psychotherapy* 30, no. 4 (2014): 517–532.
Shaddox, Karl. "Generic Considerations in Ishiguro's *Never Let Me Go*." *Human Rights Quarterly* 35, no. 2 (2013): 448–469.
Sheldon, Rebekah. *The Child to Come: Life after the Human Catastrophe*. Minneapolis: University of Minnesota Press, 2016.
Suter, Rebecca. "Untold and Unlived Lives in Kazuo Ishiguro's *Never Let Me Go*: A Response to Burkhard Niederhof." *Connotations* 21, nos. 2–3 (2011–2012): 391–406.
Tucker, Jeffrey A. "'The Human Contradiction': Identity and/as Essence in Octavia Butler's *Xenogenesis* Trilogy." *Science Fiction Studies* 37, 2 (2007): 164–181.

6 Climate Change and Apocalyptic Literature
Post-Human Stewardship in Paolo Bacigalupi's *Drowned Cities* Trilogy

Jeff Karem

> The Drowned Cities: a coastline swamped by rising sea levels and political hatreds, a place of shattered rubble and eternal gunfire. It had been a proud capital, once, and the people who inhabited its marble corridors had dominated much of the world. But now the place was barely remembered, let alone in places where civilized people gathered. The histories it had dominated, the territories it had controlled, all had been lost as its people descended into civil war – and eventually were forgotten.[1]

Since the beginning of recorded history, humans have been thinking, paradoxically, about the end of that history. Predictions of the end, along with its meaning and significance, have been a vital element of storytelling traditions throughout the world. Scholars describe this literature as "apocalyptic," a word whose Greek root (*apocalypsis*) means "to uncover" or "to disclose."[2] For most of human history, stories of the final unveiling at "end of days" have had an explicitly spiritual or religious significance. In the Christian traditions of the West, "apocalypse" signifies a final unfolding of God's plan, as elaborated in such texts of John of Patmos's Revelation or other visions of the Last Judgment. Indigenous stories throughout the world frequently have foretold an end in the form of a cleansing flood arising as a punishment for human transgressions against deities or the natural balance of creation.[3] One could rightly describe the latter category of stories as the first cautionary tales advocating for stewardship. Whether providential or punitive, these older apocalyptic traditions share the common tenet that forces far greater than humanity will be responsible for the end of the world as we know it. Although humans may trigger apocalyptic events based on their actions in these stories, they will not be the agents of the apocalypse themselves.

In the twentieth century, however, apocalyptic literature took a new turn because the tools of modern warfare gave humanity the capacity to bring about its own apocalypse, with no divine intervention required. Apocalyptic stories, while still an important thread in religious traditions, became a subgenre of science fiction, significant not only in literature, but also in cinema, television, and other categories of popular culture. In the

DOI: 10.4324/9781003219064-9

absence of a guiding religious or spiritual final plan, these stories often turned toward a contemplation of what would come after the cataclysm or disaster that ended the existing social order, enriching the genre of apocalyptic literature with a new emphasis on a "post-apocalypse" world. In spite of its speculative and futuristic emphasis, post-apocalyptic science fiction is firmly rooted in the anxieties and concerns of its cultural context. In fact, with their focus on extreme social, political, and environmental transformations, these texts provide some of the sharpest images of the fears of a particular cultural moment, often tied to concerns that humanity will not steward its knowledge and technology wisely. For example, the new and devastating weaponry in World War I (machine guns, high explosives, and chemical weapons) inspired arguably the first post-apocalyptic story in American literature.[4] The shadow of Hiroshima and Nagasaki, along with revelations of Nazi germ warfare plans, gave rise to a powerful body of Cold War literature that contemplated the aftermath of a nuclear[5] or biological holocaust.[6] With the fall of the Berlin Wall in 1989 and the demise of the Soviet Union, nuclear fears abated (though they rightly did not disappear) and there emerged, arguably, a relatively fallow period in post-apocalyptic fiction during the 1990s. Perhaps because of rapid economic growth and general optimism about the internet revolution, there are few examples of that genre in that decade, with the notable exception of apocalyptic scenarios tied specifically to artificial intelligence or robotic technology, such as the Terminator franchise in American cinema.

In the twenty-first century, however, there has been not only a renaissance of post-apocalyptic fiction, but a Golden Age, if one can say that of such a dark genre. Many of the first post-apocalyptic novels in the new century derived from post-9/11 anxiety in the West about terrorism, sleeper cells, and international threats connected to diseases like Ebola. A number of scholars have argued that fears of an invasive "other" or enemies "contaminating" the nation inspired the zombie apocalypse boom that started in the aughts, such as Max Brooks' *World War Z* (2007) and Robert Kirkman's *The Walking Dead* graphic novels (2003–2019) and television series adaptations (2010–2021).[7] Although climate change fiction (or "cli-fi," for short) is not a twenty-first-century innovation,[8] catastrophic climate change has been central to many recent post-apocalyptic narratives, and some of the most acclaimed novelists in English have made significant contributions to the genre. Margaret Atwood's *MaddAddam* trilogy (2003–2013) explores the damage humans have wrought to the earth's ecosystems through rapacious agriculture, extractive industries, carbon emissions, and unmitigated pollution.[9] Chang Rae-Lee's *On Such a Full Sea* (2014) depicts an Earth transformed by rising temperatures and rising sea levels.[10] Cautionary tales tied to climate change have also been extremely important to recent young adult fiction (or "YA" for short). Because an entire generation has grown up with the clear and present danger of cataclysmic climate change and will be living with the consequences of these developments for the rest of their lives. it should be no surprise that YA has produced some of the most resonant texts exploring a climate change apocalypse. Suzanne Collins' famed *Hunger Games* trilogy focuses on an American continent facing starvation and a brutally oppressive government due to worsening natural disasters and a collapse of natural resources.[11] Marie Lu's *Legend* series (2011–2019) also features climate change as an immanent cause for breakdown of democracy and the rise of newly fascist political and social orders.

As prescient as these texts may be in their vision of how climate change will produce scarcity and displacement, they still profess a faith that the outcomes can be managed – for good or for ill – by human agency. Put another way, the majority of these recent post-apocalyptic texts regarding climate change are anthropocentric. Part of the reason for this focus may be the intersection of post-apocalyptic fiction with dystopian narratives. Fictions of dystopia (back-formed from Thomas More's *Utopia* and translating as "bad place")[12] have a history of being intertwined with apocalyptic and post-apocalyptic fiction, although they are not always overlapping. Dystopia typically refers not simply to a troubled or grim human community, but to a highly ordered society that regiments human behavior and resources, usually for a totalitarian goal that trumps individual rights. As such, human agency, in the form of restricting other humans' freedoms, or in resistance to those restrictions, is at the heart of the dystopian genre. In most classic twentieth-century examples of dystopian fiction – such as E.M. Forster's "The Machine Stops" (1909), Aldous Huxley's *Brave New World* (1932), George Orwell's *1984* (1949), and Ray Bradbury's *Fahrenheit 451* (1953) – humans subordinate others for the sake of social control, maintaining a political regime, or sustaining a war against an opposing nation. In recent climate change fiction, dystopian regimes often arise as a way to assert order and manage resources in world afflicted by scarcity and civil unrest because of a changed climate. In Margaret Atwood's *Oryx and Crake* (2003) and Chang-Rae Lee's *On Such a Full Sea,* highly-ordered corporate city-states arise as means of husbanding resources and protecting against class uprisings in a climate-disrupted world. In *The Hunger Games* trilogy, an overarching elite known as the Capitol governs the twelve Districts with nearly absolute power by control of the food supply, brutal military interventions, and the demand of annual fights to the death among tributes from the districts. The protagonists in Marie Lu's *Legend* series face a variety of forms of military and political control in a North America fragmented into warring, often totalitarian, governments.

While humans in these climate fiction dystopias are responsible for climate change and repression, humans are also able to resist those regimes and to restore political and environmental balance. Katniss brings down the Capitol in Collins's *Mockingjay* (2014); June and Day bridge class divides in Marie Lu's novels to lead a rebellion against the corrupt Republic. For every problem created by humanity in Atwood's *MaddAddam* trilogy, there is a human hero or anti-hero vested with corrective power. In *Year of the Flood* (2009), Atwood presents a sect called "God's Gardeners," which takes upon themselves the explicit task of stewardship neglected by the ascendant corporate-states.[13] The Gardeners, even in the midst of urban settings like Toronto, effectively "opt out" of the modern economy by developing their own hydroponic gardens, rooftop beehives, and herbal remedies. Even Crake, the villain of the trilogy who precipitates the "Flood" by engineering a deadly virus, understands his work as stewardship – curbing humanity's excesses by dwarfing their numbers.[14] After Crake's virus-induced apocalypse, the remaining God's Gardeners join together with other survivors to promote a new social structure that respects and preserves creation, effectively securing a more reciprocal partnership with nature as the final triumph of the trilogy. As one of the storytellers advisers his listeners in *MaddAddam*, they have "cleared away the bad men... to make a good and safe place for us to live."[15] These richly biblical allusions in the trilogy – Adam, the Garden, the Flood – suggest that, for all of its darkly predictive vision, there is hope for a new Eden to emerge from the apocalypse, on the basis of humanity finally embracing the stewardship responsibilities that it was supposed to practice from the beginning. As Deborah C. Bowen notes, even the small group of posthumans in the novel, the Crakers, become more like humans by the end of the trilogy "by acquiring a mythology and a religion... by acquiring the beginnings of writing and history; by fostering the ability to think ahead and to plan."[16] In sum, Atwood's vision of stewardship is ultimately anthropocentric. Humans abused their power and ruptured creation, but humans can make it right again, both by changing their behavior and by leading other species to a more enlightened path.

In contrast to the faith in Collins, Lu, and Atwood that humans, for better or for worse, will decide the fate of the world, Paolo Bacigalupi's *Drowned Cities* trilogy (2010–2017) advances a vision of stewardship and the earth that radically challenges anthropocentricism. In the three novels of this series – *Shipbreaker* (2010), *Drowned Cities* (2012), and *Tool of War* (2017) – Bacigalupi upends expected norms of global and environmental stewardship by reframing climate change as a corrective response to human excesses. In a grim fulfilment of the worst-case scenarios of predicted by contemporary oceanographic experts,[17] rising sea levels in these novels have overwhelmed the vast majority of cities, decimating the old economic and political economic order throughout the world. Bacigalupi's fiction is worthy of critical observation not only for its stark inversion of the expected paradigm of human stewardship, but also for its contemplation of many other "posts" beyond the apocalypse. Although the trilogy is set in North America, there is no longer a United States government, and the remaining humans have essentially grouped into small, local tribes or city-states, which are often at war with one another. The electrical and telecommunications grids have fallen, and petroleum reserves are gone. Bacigalupi's world is decidedly post-national (unlike, for example, David Brin's *The Postman* (1986), in which the hero aims to create a Restored United States of America in the wake of a global war between the United States and the Soviet Union). While there are hints that China may still have a national government (though we never see it in the novels), global power is projected primarily by multinational corporations who compete for the remaining resources. Very few people are part of that corporate elite, however, and there appears to be nothing like a middle or working class. The vast majority of characters make a living either via subsistence agriculture or scavenging materials for re-sale. Scavenge work gives the title to the first volume, *Shipbreaker*, which focuses on a young man named Nailer, who works as a part of Gulf Coast crew that dismantles abandoned tanker ships and sells their part for scrap. Scavenging is at the base of the corporate economy because humanity has exhausted the earth's mineral and metal resources through its extractive approaches to agribusiness, mining, and industry.

Rather than affirming the power of human agency to save the earth, Bacigalupi's novels demonstrate that the earth does not need saving – by us. His fiction proposes that the earth, as a global system, will protect itself by purging itself of organisms with unsustainable ways of life, which includes the majority of twenty-first-century humanity. By Bacigalupi's logic, humanity needs to practice stewardship to protect its own survival as a species, not because the planet needs our wisdom, for it will endure – and likely flourish – after the end of the Anthropocene. In terms of politics and economics, Bacigalupi's novels suggest that dystopia is not a post-apocalyptic scenario in the future but the core of humanity's present in the twenty-first century. While the world is significantly transformed by increased temperatures, rising sea levels, and attendant climate change, it is humanity, not the earth, which is in jeopardy in his fiction. In keeping with this spirt, the protagonists across the trilogy lack the agency for redemption of the fallen world: their goal is generally survival. The trio of young people at the heart of *Shipbreaker* – Nailer, Nita, and Pima – seek to evade Nailer's father and corporate agents who want to ransom Nita to her wealthy father and kill the other two for helping her escape capture. Mahlia and Mouse, in *Drowned Cities*, aim to survive a civil war raging in what used to be Washington, DC. The title character of *Tool of War*, a half-human, half-animal super-soldier escapes enslavement as a war-beast and fights to preserve his life – and his freedom. With Tool, the augmented half-man (or "augment," for short) linking all three novels, Bacigalupi introduces perhaps his most challenging "post" to the reader: his vision of a post-human future. This idea resonates broadly across his trilogy because humanity is no longer the dominant species in most of North America. Augments like Tool and his kin are post-human hybrids, and while they are enslaved at the start of the trilogy, they are poised to become ascendant because they are better equipped to adapt to the new ecosystems that have developed from climate change. Although augments have been genetically conditioned to be obedient to the humans who purchase them, Tool has broken that conditioning, and he advises his human friends not to underestimate augments: "Do not be so certain of what my kind can and cannot do. We are faster, stronger, and whatever you may think, smarter than our patrons."[18] While Tool and his brethren were created primarily for labor and war, Tool becomes much more than that – a visionary, a leader of men and hybrids, and a chorus-like figure who reminds the reader of humanity's failures and prophesies that the arc of history will bend towards the post-human because of our poor stewardship.

Bacigalupi effects this challenging vision through a series of recurring motifs and varied techniques. Through a wide range of imagery, he reminds readers of humans' animality and their precarious place in the natural ecosystem. Bacigalupi likens humans in the new ecosystem to scavenging insects:

> Wherever the huge ships lay, scavenge groups like Nailer's swarmed like flies. Chewing away at iron meat and bones. Dragging the old world's flesh up to the beach to scrap-weighing scales and the recycling smelters than burned 24-7 for the profit of Lawson and Carlson, the company that made all the cash from the blood and sweat of the shipbreakers.[19]

Such scenes of scavenging evoke a powerful sense not only of the diminished scale of humanity, but of the obsolescence of the "old world" in its reliance on oil and heavy machinery. Bacigalupi compares the old hulks to "rusting dinosaurs" and "great wallowing brutes," evoking images of extinction and fossilization.[20] The description of corporations' turning a profit "from the blood and sweat of the shipbreakers" points out that the shipbreaker economy feeds not simply on recycled materials, but on the human resources of the shipbreakers themselves. As Johan Höglund notes, "Nailer is being eaten by a completely unregulated and ruthless economy that understands human bodies only as a (cheap) labour resource."[21] Bacigalupi prophesies an even more literally consuming side to this scavenge economy with the literal use of humans as a resource. An unemployed shipbreaker, Sloth, faces the prospect of selling some of her organs to Harvesters, "medical buyers [who] can slice and dice her like a side of pork." When the "swank" (upper-class) Nita is kidnapped by Nailer's father, he grimly predicts the outcome if a ransom is not paid: "Be tragic if we ended up scavenging our rich girl for spare parts, wouldn't it?"²² The phenomenon of humans preying on other humans echo the biological concept of predator cannibalism, in which a predator that has over-preyed its own food source will be forced to prey on its own.²³ Bacigalupi's depiction of intra-specific predation, coupled with a scavenging economy, suggests an exhausted species that has fallen from the apex of the planet's ecosystem. In *Drowned Cities*, which focuses on the interior of the American continent, the reader finds a wild landscape full of predators who block travel and prey upon humans. Alligators have spread north all the way to the Mason–Dixon line and grown enormous in the flooded and warm landscape near rivers and coasts. Endangered species like the Florida panther have resurged and now scale abandoned dwellings to take humans in the night. In a futuristic echo of Euro-American folklore, coyote–wolf hybrids (coywolves) dominate the continent's forests and cut human settlements off from one another.²⁴ Even soldiers embroiled in a local civil war near the Chesapeake Bay find themselves vulnerable: "They were hunters. But now, as night closed in on them, and the swamp became black and hot and close, they were becoming prey."²⁵ In *Drowned Cities*, when Mahlia contemplates crossing a swamp to rescue her friend Mouse from those same soldiers, Tool asks her, "You think you are some fine predator? A swamp panther or coywolv?... Where are your teeth and claws, girl?... "Where is your bite?"²⁶ Because Mahlia helps heal the fugitive Tool, he helps her make the swamp crossing, but his litany of rhetorical questions nonetheless stands as a stark reminder of humans' precarious place in the wilderness.

In each novel of the trilogy, Bacigalupi depicts not only the specific challenges faced by his protagonists in this warmed and flooded ecosystem, but also the radical fall of human civilization in the face of these changes. He unveils this broader portrait by examining a formerly great American city as a focal point in each of the three novels. His choice of cities is revealing, and each forms a chronotope²⁷ embodying crucial spaces and moments in American history. New Orleans is the hub of *Shipbreaker*; Washington, DC is the center of conflict in *Drowned Cities*; Boston is at the heart of *Tool of War*. The geographic trajectory across these novels is significant in itself, as it follows the arc of Tool escaping his enslavers and seeking freedom – a clear echo of the journey north of fugitives in nineteenth-century American literature. All of these cities are, to varying extents, drowned, and Bacigalupi's careful portraits of each foretell an ascendant post-human ecosystem coupled with human decline in general, and American decline specifically.

Welcome to the Jungle: The Kudzu and Voodoo of Orleans

New Orleans forms an especially apt starting point for the *Drowned Cities* trilogy. New Orleans is one of North America's most politically and economically significant cities. At the juncture of the Mississippi

River and the Gulf of Mexico, it has connected shipping and trade from the American continent to the Caribbean, the Atlantic, Europe, and Africa. Its own history is a palimpsest, with overlapping layers of colonial dominion, as it has been a key port for Spanish, French, and U.S. empires. In the context of Bacigalupi's examination of economic exploitation, New Orleans resonates because of its role as a critical nexus in the trans-Atlantic slave trade – a history that echoes in its service as port for trafficking augments like Tool in *Shipbreaker*.[28] New Orleans also has extremely contemporary resonance in discussions about stewardship and climate change. Published in 2010, *Shipbreaker* was written in the shadow of Hurricane Katrina, a super-storm event that decimated the city and forced thousands of U.S. citizens – primarily African Americans – to become climate refugees within their own country.

Before the reader encounters Orleans (denizens have dropped the "new" from its name) in *Shipbreaker*, Bacigalupi confronts readers with a Katrina-like storm that destroys the shipbreakers' Gulf Coast beach settlement. These storm scenes in the novel are rich in specific detail and metaphorically complex. Personified as "city killers" (a new class of hurricane), the storms are represented with fearsome agency. The first storm in the novel "slashe[s] the beach" with its rain; as the final one in the novel gathers, it "roil[s] like a seething cauldron of snakes."[29] These figurations invest the storm with a power that is not only super-human, but also mythically resonant, as if the storm were an avenging swordsman or a witch seeking retribution with a powerful spell. Bacigalupi relies more on science than mythology to elucidate the origins of the city-killers, however, as they are directly tied to failures of human stewardship. In a bitter double irony, oil drilling in the gulf has not only released carbon that increased global temperatures and thus storms' intensity, also but wrecked the marine ecosystems that have historically served as a buffer against hurricanes. Nailer's friend Nita explains the viciousness of petroleum exploration:

> They got it [oil] from everywhere... From the far side of the world. From the bottoms of the sea... They used to drill out there, too, in the Gulf. Cut up the islands. It's why our city killers are so bad. There used to be barrier islands, but they cut them up for their gas drilling.[30]

As terrifying as the storm is, and as devastating as its effects are for the shipbreakers (almost all of their settlement is destroyed), Bacigalupi shows the beauty of its aftermath, as it appears to have cleansed the beach: "The beach was empty. Not a sign of human habitation... The soot was gone, the oil in the waters, everything shone brightly in the blaze of the morning tropical sun." Pima comments with wonder to Nailer, "It's so blue. I don't think I've ever seen the water so blue," but Nailer cannot answer: "Nailer couldn't speak, the beach was cleaner than he'd ever seen in his life."[31] While it would an overstatement to suggest that a single storm has restored an ecosystem to a pure state, the bright colors and luminous beauty suggest the power of nature to restore itself in the fullness of time, de-centering the place of humans and checking their privilege. One might say, rephrasing the title of George R. Stewart's classic post-apocalyptic novel, that "the sea abides" in Bacigalupi's novels.

The resilience of nature is revealed in greater grandeur as the novel approaches New Orleans. Nailer and his friends flee their settlement in the wake of the storm and to avoid pursuit by Nailer's father and corporate kidnappers. They hitch a ride on a train in the hope of booking passage out of the Gulf at Orleans and the find that the vehicle barely makes a mark on the landscape: "The twin rails of the train tracks were being swallowed by the dense jungle." As they approach the city, the reader encounters Bacigalupi's first tableau of the dramatic changes that rising sea levels and expanding forests have wrought on the human metropolis:

> The great drowned city of New Orleans didn't come all at once, it came in portions: the sagging back of shacks ripped open by banyan trees and cypress. Crumbling edges of concrete and brick undermined by sinkholes. Kudzu-swamped clusters of old abandoned buildings shadowed under the loom of swamp trees... They sped above the mossy broke-back structures of a dead city. A whole water-logged world of optimism, torn down by the patient work of nature. Nailer wondered at the people who had inhabited those collapsing buildings. Wondered where they had gone. Their buildings were huge, larger than anything in his experience at the ship-breaking yards. The good ones were built with glass and concrete, and they'd died just the same as the bad ones that seemed simply to have melted in on themselves, leaving rotting timbers and boards that were warped and molded and sagging.[32]

This description strongly echoes the aftermath of Katrina in New Orleans, with a notable exception in terms of racial and class politics and natural disasters. The devastation wrought by Katrina was not distributed equally, as the predominantly African American wards suffered far more than upper-class neighborhoods due to inequitable infrastructure and gaps in federal assistance. As Nicole Waligora-Davis describes it, "With the limited resources, the poor in New Orleans were less able to evacuate before the storm. In a city where poor and black were so desperately intertwined, it was these citizens who suffered Katrina's most grievous effects."[33] In Bacigalupi's Orleans, by contrast, the storms and rising seas have been so intense that "the good [buildings]... died just the same as the bad ones," suggesting the that the storms have leveled both physical and class structures in Orleans. When Pima and Nailer ask Tool why the wealthy and powerful could not protect even their own parts of the city, he answers laconically that "[t]hey did not anticipate well" – a pithy description that aptly encapsulates the failures of human stewardship in the face of climate change.[34] In fact, the only persons in Bacigalupi's Orleans living above a bare subsistence level are visitors or traders docked in ships outside of it. Everyone else is, to varying degrees, a laborer, servant, or scavenger serving these outsiders. Perhaps as an ironic play on anthropocentricism, Bacigalupi renders the city with subtle humanizing imagery. The buildings "died" and the city is "broke-back." Rather than elevating the city, these personifications evoke its fragility. What mortal body can stand firm in comparison to the work of nature? Bacigalupi's vista of the city in the novel extends this personification further in a grim direction – "The dead city, still half alive, like a zombie corpse reanimated"[35] – skillfully evoking the voodoo history of the city as a way to explain its liminal state, with a notable difference. In voodoo tradition, someone else must make you into a zombie; New Orleans' zombification here is self-inflicted because of poor human stewardship of the city and the natural environment.

From Sea to Rising Sea: Washington, DC

The denouement of *Shipbreaker* wraps up the plots of the human characters, leaving Tool as the only character continuing into the next novel. Nita, Nailer, and Pima escape their pursuers with the help of a clipper ship loyal to her father, Jayant Patel, head of Patel Corporation International. By the start of *Drowned Cities,* Tool has journeyed North to Washington, DC to seek revenge upon his old enslaver and guarantee his freedom. The importance of the capital city as an emblem of American power is obvious, but several features of the city's history are relevant for appreciating Bacigalupi's rendition of it. First, in the context of Tool's emancipation plot, it is vital to recall that Washington, DC was built by slaves and permitted slavery until 1862. In *Drowned Cities*, slavery has arisen again as the warring factions in the city force their captives into scavenge labor so that the militias can sell resources for weapons. This second novel of the trilogy is literally a story of civil war among violent factions, so Washington, DC is an ideal locus for Bacigalupi's speculative vision. Significantly, there is no Union versus Confederacy binary in Bacigalupi's Washington. No one opposes slavery except for Tool and those who join his cause, and his freedom is in jeopardy at the start of the novel because he finds himself captured by a militia. The entire city is riven by a multiplicity of factions that, depending upon one's perspective, echo either the paramilitary violence of a failed nation state, or the contemporary tribal politics in twentieth-century Washington, DC. Before Bacigalupi offers a broader view of the city, he provides a portrait of the political and economic landscape, and it is even more dire than that in *Shipbreaker*. He frames the civil conflict with strong animal imagery: "the Drowned Cities were full of fighting factions, perpetually tearing at one another's throats." This figuration levels the differences between humans and animals, suggesting that the partisans are fighting packs comparable to the coywolves outside the city. In fact, Tool makes the case to Mahlia that humans, in all of their partisan division, are worse than animals for their fratricidal violence: "Your kind has always been garbage. Willing to run when you should stand. Willing to kill one another for nothing other than scraps. Your kind... worse than hyenas. Lower than rust."[36] As in *Shipbreaker*, the economy is based on scavenge, but the fact that the setting is the U.S. capital makes the effects more dramatic for the reader. Rather than harvesting parts from wrecked ships, the soldiers and their slaves in Washington, DC are taking apart the infrastructure of the capital itself, "picking over history's bones"[37] – an unsettling image of a capital corpse, with humans as vultures. One of the protagonists, Mahlia, reveals to Tool as they approach the city that her mother, when she was still alive, had made a living selling antiquities from the capital to foreign peacekeepers before they abandoned the continent as a lost cause. Mahlia continues this tradition at the end of the novel when, in order to book a ship to escape the city, she proposes to sell a flag to a foreign trader. She tells her friend Mouse, "Put up that old flag. The one with the stars in a circle, and the red and white stripes."[38] At first glance, the image of Mahlia flying an American flag might seem an emblem of national resilience, but the fact that Mahlia cannot even recognize the flag and has to describe it by shapes and colors, suggests that the United States is a concept totally alien to those in Bacigalupi's world. The fact that she is flying the Betsy Ross thirteen-star flag to advertise her willingness to sell it provides a grim image of commodification. In the Drowned Cities, America has become Americana.

When Bacigalupi's narrative focus pulls back to examine the cityscape itself, the vision of decline is even more pronounced. When Mahlia scales an abandoned building in the DC suburbs, she witnesses how water and the jungle have transformed the city:

> Five stories up, the jungle spread in all directions, broken only where the war-shattered ruins of the Drowned Cities poked higher than the trees. Old concrete highway overpasses arched above the jungle like the coils of giant sea serpents, their backs fuzzy and covered, dripping long tangled vines of kudzu... Rectangular green pools pocked the fields in regular lines, marking where ancient neighborhoods had once stood, the outlines of basements, now filled with rainwater and stocked with fish. They glittered like mirrors in the hot sun, dotted with lily pads, the graves of suburbia, laid open and waterlogged.[39]

This imagery echoes and intensifies the representation of the cityscape in *Shipbreaker*. Kudzu and water are again the victors, and the vista closes with a macabre image. Whereas Orleans was a zombie, here the suburbs are likened to flooded and open graves.

As in *Shipbreaker*, Bacigalupi renders cities with personification, and the conjunction does not flatter either the city or the humans who built it. When Mahlia gets closer to Washington, DC, the ruins form a fearsome sight: "The buildings rose up, like bodies staggering up out of the grave. Towers and warehouses and glass and rubble. Piles of concrete and brick where whole buildings had collapsed."[40] The corpse imagery is especially appropriate because the capital is filled with the bodies of the dead who have died in the civil war, and the buildings have been leveled not only by nature, but also by artillery and bombs. In this light, the devastation in DC reveals an even worse landscape than in *Shipbreaker*. Nature has a much easier time retaking the city because humans are already tearing it apart themselves. With a more subtle tenor, Mahlia's friend Mouse looks at the city and sees an emblem of aging humanity:

> From a distance, if you didn't listen for the warfare, the place could have been abandoned. But as you got closer, you could make out the details. Trees sprouting from windows, like hair from an old man's ears. Robes of vines draping off sloped shoulders. Birds flying in and out of upper stories. [H]e surveyed miles of ancient buildings and swamped streets turned into canals. Networks of algae-clogged waterways were dotted with lily pads and the stalks of white lotus flowers. Block after block of buildings were swallowed up to their second story and sometimes higher, like the whole city had suddenly decided to wander off and go wading in the ocean.[41]

These images of "hair from an old man's ears" and "sloped shoulders" evoke the impression of an elderly relative in declining health. In contrast, the other flora and fauna are vibrant, abundant, and adaptive. Birds that may have collided with glass when the buildings were intact can now navigate in and among them with ease. The closing image intensifies the sense of decline with a figuration of the city as an old man losing his bearings and stumbling into the sea, presaging dotage and dementia for the center of the American power.

The conclusion of *Drowned Cities* makes clear that Bacigalupi aims to compound his critique of human dominion in general with a specific evocation of the contradictions of the United States. Tool is recaptured by a former enslaver whose base of operations is the capital building itself. He chains Tool to the marble to show rival factions that he is master of man and half-man, but his plan fails when Tool, like Samson, tears down the pillars to which he is chained. Tool vows to bring an end to the civil war and restore order to the city, recognizing the difficulty of the task. He tells

Mahlia and Mouse that it will always be difficult to bring peace among humans because

> [i]t only takes a few politicians to stoke division or a few demagogues encouraging hatred to set your kind upon one another. And then before you know it, you have a whole nation biting its own tail, going round and round until there is nothing left but the snapping of teeth... I have never seen a creature so willing to rip out its neighbor's throat.[42]

This passage – one of Tool's closing comments in *Drowned Cities* – is rich with ecological and political implications. The images of snapping teeth and ripped throats reduce humanity to a bestial level of violence, with an acerbic figuration of the nation embodied as a foolish dog chasing its tail. The final comment is particularly lacerating coming from Tool, who values pack loyalty and finds it abundant in the wild predators of the novel, but conspicuously absent in humans. Lastly, the image of harming one's neighbor seems a clear subversion of the putative claims of politicians that the United States has been a Christian nation. The descendants of America in the Drowned Cities do not "love thy neighbor" but instead rip out their throats. Written four years before the rise of Trump and nine years before the Capital insurrection, Tool's words are highly prescient in anticipating the divisions that Bacigalupi's readers would later see in the nation's capital, even before the apocalyptic transformations in his trilogy.

Cradle of Liberty No More: Seascape Boston and the Hunt for Tool

Tool of War, the concluding volume of the trilogy, begins with Bacigalupi offering a kind of memento mori for the U.S. capital, where Tool claims victory over the last militia general:

> The palace he stood in was a ruin. Once it had been grand, marble floors, majestic columns, ancient masterful oil paintings, a graceful rotunda. Now he stood under a shattered dome, and could survey the city he warred for, thanks to a bombed-out wall. He could see right out to the ocean where it lapped below, on his very front steps. Rain spattered in and made thin, slippery pools on the floors. Torches guttered in the damp, giving light for the human beings, so that they could see the barest edges of what Tool could see without any aid at all. A tragic ruin, and a site of triumph.[43]

Tool's triumph proves to be short-lived. He vows to rebuild the city now that he has brought peace, but he comes to the grim recognition that the humans are much better at destroying then building. Put simply, no one in the city has skills beyond fighting or scavenging. Frustrated in discussing plans with one of his soldiers, "Tool wanted to howl in his face. *Make your own way! Build your own world! Your kind constructed me! Why must I construct you?*"[44] Tool's outburst well captures the pivot that Bacigalupi makes in the final novel of the trilogy. While *Shipbreaker* and *Drowned Cities* follow humans and Tool through a broad portrait of a flooded North American landscape and seascape, *Tool of War* has a more intimate focus on Tool's origins as a genetically engineered hybrid, his quest for self-ownership, and the implications of the presence of increasingly autonomous augments for the world of humans. With the exception of a brief description of the history of Seascape Boston, *Tool of War* does not offer vistas of drowned cities and foregrounds instead Tool's interiority, dreams, and visions. This internal exploration runs parallel to a physical flight to freedom necessitated when General Caroa, Tool's creator and former enslaver, destroys Tool's headquarters and his entire army with a drone strike.

Severely wounded, Tool swims away from the firestorm and clings to a ship heading north to Seascape Boston, echoing the journey of past fugitives like Frederick Douglass. Fortunately for Tool, Mahlia, the human protagonist he had helped in *Drowned Cities*, is also on the ship. At first glance, their arrival seems a moment of triumph, but Bacigalupi subverts that expectation in multiple directions. The inner sections of Seascape Boston have escaped storms and flooding, so those parts of the city are still functional and house a major shipping corporation, Patel Global. The reader learns, however, that the city could have saved much more if its leaders had listened to the scientists warning of rising sea levels. Instead of respecting its scientists, the city jailed a leading oceanographer for vandalism when he marked the impending new sea level throughout the city as a warning to its residents. When he is arrested, he tells the city leaders, "People don't mind that the sea will swallow their homes, but woe to the man who paints their future for them."[45] When a Category Six hurricane levels much of the city, the scientist and his family develop a plan to build seawalls from the wreckage. Seascape Boston marks an example of rare resilience in Bacigalupi's trilogy, but it also marks another example of failed stewardship. The city's inhabitants refuse to take any proactive action and respond only after there is a devastating storm. As Tool reflects to himself about humans' failures to plan, "*These storms are your creations. You made them. Now you struggle to survive.*"[46]

Although parts of Seascape Boston literally are elevated above drowning, the city fails to maintain any kind of moral high ground. Bacigalupi reveals that the city is like any other in its exploitation of augments, and it is most definitely not a beacon of liberty for a fugitive like Tool. In fact, the challenges Tool faces in Seascape Boston form a grim echo of nineteenth-century history in the United States. The operatives from Mercier, the corporation who engineered him, infiltrate the city and chase Tool like slavecatchers in the antebellum United States. When Tool seeks help from Nita Patel, the company heir he helped in *Shipbreaker*, he receives medical aid but learns that her company is legally obligated to return him to Mercier Corporation, much as nineteenth-century U.S. citizens were bound by the Fugitive Slave Acts. Nita's father rebuffs Tool's entreaties by reminding him of the law: "Mercier demands you be returned to them, alive are dead, and they are more within their rights."[47] Significantly, it is only in the context of the claims of human enslavers that the phrase "right" arises in the novel. In this respect, Seascape Boston's moral landscape may actually represent a devolution in comparison to United States history in terms of dominion and abuse. While Douglass and Jacobs found allies in abolitionists like William Lloyd Garrison or Lydia Maria Child, there is no group of humans to offer legal or political support to Tool, with the exception of the small crew made of those he befriended in the two prior novels. Taken together, these human failures in Seascape Boston suggest that rather than learning from their mistakes of past domination of others, humans will continue this pattern in the future, for as long as they can. Understanding this pessimistic conclusion is vital for appreciating how Bacigalupi's critiques of human stewardship intersect across the trilogy. Writ large, humanity's extractive approach to the earth depleted vital resources, destroyed ecosystems, and accelerated global climate change. The exploitation of augments as a resource is part and parcel of this sense of dominion. In *Tool of War*, Bacigalupi predicts an end to that dominion, with the suggestion that humans' abuse of augments may be as equally disastrous for humanity as their assertion of mastery over the earth more broadly.

From Revolution to Evolution

Although Bacigalupi's rendering of Seascape Boston highlights its failure to honor its history of protecting fugitives, his portrait does suggest another kind of historical resonance: a site of revolution. While the city fails to extend a hand of liberty to Tool, it is there that he asserts his sovereignty and aims to form a coalition with other augments. Tool begins this process with a challenging moment of personal defiance to the head of Patel Global. With rhetorical flourish befitting a classic orator, Tool rebuts the claims of ownership by his own embodiment and presence:

> Because they call me property? Because some documents claim I am their thing? I'm sure they have many documents making such claims. I'm sure they say they own my blood design and my genetic mix. That I am *intellectual property*, from head to toe, from fang to claw. And yet here I remain… and still I do not obey.[48]

Tool embodies a Miltonic exemplum of *non serviam* with a righteous twist. While Milton's Lucifer disobeys God because of his own arrogant self-conception, Tool simply asserts that he has a self and agency, revealing the hubris and overreach of his creators.

Tool's recognition of his revolutionary potential emerges primarily in a series of memories, dreams, and visions in which he recognizes the common connections he shares with augments – even those he was tasked to fight against from rival corporations. On the basis of a common origin, he understands that they share a common cause if they can awaken to it. When he sees fellow augments in Seascape Boston, he feels a powerful kinship and wants to reach out to speak to them:

> Tool felt a vibration of connection with them. Tool found himself leaning forward, filled with an almost desperate desire to see him as a brother. *Are we not all molded from the same clay? All knitted together from the same strands of science? Do you not see that we are one? We are brothers!*[49]

Bacigalupi demonstrates that Tool's connections to other augments extend deeper into his past. Tool recovers a suppressed traumatic memory that reveals that he is not alone in asserting his sovereignty and breaking free of human domination. When he is hiding out in the city, he starts to feel a surprising sense of connection to his sworn and greatest augment opponent, the First Claw of Kolkata, whom Tool believes he defeated in hand-to-hand combat on behalf of Mercier Corporation before he had broken free of his conditioning. Over a series of dreams that gradually are revealed as flashbacks, Bacigalupi reveals that there was once a ceasefire and a parley between Tool and the First Claw, in which they discussed their shared origins and reached an agreement not to fight each other:

> He remembered the First Claw of Kolkata, clasping his hand in bargain. *My brother*. Across the great divide of genetics and language and design and culture, they had been brothers. Across the chasm of military stalemate, monomolecular razor wire, and muddy defensive trenches, they had reached an agreement. Beneath the shining arc of mortars launched against each other, they had been *Kin*.[50]

This passage is striking not only for the sense of solidarity between First Claw and Tool, but also for the sharp contrast between the peacemaking augments and the extensive technologies of war crafted by their enslavers. Bacigalupi's roster of the warcraft surrounding the augments forms a skillful list of something old, something new, as it ranges from trenches and mortars to futuristic nanotechnology.

Bacigalupi's vision of these warriors rejecting human violence and ending battle among themselves forms an inspiring vision of a post-human future, but it is quickly destroyed by human generals, who firebomb all of the augments for their disobedience: "And because they had discovered their brotherhood, and because this was more memory than dream, Tool was saddened to know that fire would soon rain down from their angry, frightened creators, and they would die." The recompense for recovering and re-living this painful memory is a recognition for Tool that he is not alone in defying his enslavers. Other augments have the potential to do so as well. Fittingly, Tool concludes this sequence of memories with a crucial moment of recognition – "*I am awake*" – which invites the question: can Tool awaken other augments as well?[51] Bacigalupi reveals a complex answer to this question, and tracing the unfolding of Tool's ability to persuade other augments suggests the author's sophisticated vision of the potential of augments to outpace their creators not only in war, but in diplomacy as well. Tool's first moment of gaining augment allies occurs when the head of Patel Global orders his own augments to seize Tool so that he can be returned to Mercier. Tool shouts a command at them that resonates like a "thunderclap" and stops his fellow augments in their tracks. They then tell Patel, "We cannot attack our kin," but reassure the humans that "[h]e will do you no harm. His oath is good. He is our brother." Tool's command has an almost spell-like effect over his fellow augments, and he explains to Nita Patel that this power was programmed into him by General Caroa for the sake of battle: "I was designed to not just lead my own kind into battle, but to exert my influence over those I fought. To bring them over to the side of my masters. Everywhere I go, I encourage defections."[52] Bacigalupi marks a significant difference in how Tool uses his rediscovered power. Rather than weaponizing his persuasive power as his enslavers intended, whether against humans or augments, Tool exercises restraint. Tool does not command fellow augments to attack humans or commit violence, but simply to refrain from harming him. Reciprocally, both he and the augments commit not to harming the humans of Patel. Tool has, in effect, shifted the parley he established with the First Claw in his past into an inter-species détente with this group of humans.

On the basis of this power, Mercier Corporation labels Tool a "walking rebellion,"[53] and the question of where his rebellion could lead forms the critical denouement for the novel. General Caroa confides to a colleague that "I fear we will bear witness to humanity's extinction." The head of Patel Global accuses Tool of planning genocide. Tool's rebuttal makes a powerful intersectional connection between the abuse of animals and the abuse of augments. Referring both to corporate battlefields and to the savannas of Africa and Asia, Tool argues,

> Genocide? I've done nothing to cause extinction of you or yours. Look to Mercier who has wiped out every one of my kind from the face of the earth if you wish to speak of genocide. On every continent I served, they put my kin to the sword. Do not speak to me of genocide.

Bacigalupi's connection here is subtle but clear: Tool claims kinship not only with the augments slain by the corporations, but also animal ancestors of Africa and Asia (lions, tigers, leopards, cheetahs) endangered by human violence. Tool does confront Patel with the idea that augments are "the next step in evolution,"[54] but Bacigalpi does not suggest that that necessitates a genocide by augments against humans. In fact, Tool's restraint and his fellow augments' respect for oaths old and new, suggest a culture of honor and mutual respect superior to the human cultures evident elsewhere in Bacigalupi's trilogy. Indeed, humans' failures as stewards, as demonstrated throughout the trilogy, seem the greatest threat to their future.

Bacigalupi suggests that a deepening of respect between humans and nature, while too late to reverse climate change and the drowning of cities, may at least create a more just and survivable world for humans and augments alike. In parallel to Tool's liberating augments from their genetic conditioning, his assertion of self-ownership awakens some of the humans around him to a new respect for his sovereignty and individual rights. Nita, whom Tool helped in *Shipbreaker*, starts to question her own relationship with the family's augments:

> Talon. Another augment.
> Was he family?
> A friend?
> A slave?[55]

The careful, staccato cadence of these lines suggests the tortuous path Nita's conscience faces in reckoning with her family's reliance on augment labor. She ultimately has a Damascus-like conversion after failing to rationalize the caste structure between augments and humans:

> *We treat ours well*, she thought, but it was cold comfort. All her life she'd been surrounded by them. They were designed and trained to mesh with her family, her company, to do the tasks that natural human beings could not. She had never thought of them as anything other than a natural extension of her life, and the success of Patel Global. Now she couldn't help feeling there was something wrong with the very language used to describe augments. Words like *ownership* came easily when a creature was grown from handpicked cells, raised in a crèche, and was purchased from a selection of other augments. And yet, they were not identical. They had feelings. They wept at loss. Delighted in success. They were people.
> Except they weren't.
> *They are better than people*, a dark voice whispered in her mind, one that sounded a little too much like Tool. *They are the end of people.*[56]

This passage marks perhaps the best articulation of the intersection of language, power, and human dominion (and its potential end) in the *Drowned Cities* Trilogy. Bacigalupi has crafted Nita's interior monologue with rich layers of historical resonance. Her comment "*[w]e treat ours well*" is almost a verbatim repetition of enslavers' rationalizations and self-justifying language in the Antebellum United States. Anthropocentrism is perfectly distilled in her assumption that augments were "a natural extension of her life" – a sentiment that seems easily applicable to how most of humanity treats nature, whether in Drowned Cities or in our own world. The passage concludes with both a moral recognition of the depth of augment rights and a haunting suggestion that humans have forged their own extinction by engineering an organism that excels in the brave new world of climate change in a way that humans cannot.

The prospect of human extinction – whether from climate change or at the hands of augments – haunts the final novel of the trilogy. This possibility echoes the fears enslavers had of African American rebellion in the antebellum United States, a historical feature that will resonate strongly in the final battle in *Tool of War*. In fact, Tool recognizes this Darwinian calculus explicitly when he considers the future of humanity: "Such was the way of evolution and competition. One species replaced one another in the blink of an eye. One evolved; one died out... *Some species are meant to lose.*"[57] For his part, Tool opts not to advance that extinction, in no small part because he develops a respect for the capacity of some humans to see beyond an anthropocentric perspective. Although Tool lacks any legal status or support in these novels, he does find support from humans like Nailer, Nita, Mahlia, and Mouse, whom he terms his "pack," with mutual loyalty among them forged in adversity. Reflecting upon the humans who have helped him flee the authorities, Tool "felt a surprising rush of camaraderie for these humans who dedicated themselves to his survival."[58] Bacigalupi demonstrates that Tool's respect also extends beyond simply helping his allies in battle. When Mahlia is wounded by Mercier forces trying to recapture Tool, he helps her escape and works to restore her. In a moving scene, he comments tenderly to Mahlia that "your kind is fragile," and then gives her an infusion of his own blood ("My blood will help you heal"), which not only sustains her but accelerates her recovery because of the self-repairing nanotechnology in his augment blood.[59] With his awakening from near-death, his healing through his blood, and his capacity to usher in the end-of-days for humanity, it does not seem an overreach to interpret Tool as a Christ-like figure in the trilogy. In such a reading, Bacigalupi has powerfully decentered Christian tradition, however. Rather than affirming the holy mystery of god-in-man from the New Testament, Bacigalupi links the animal to human, with no sense of the divine, especially in the bulk of humanity in his fiction.

The tense question of what Tool's leadership will unleash in the world dominates the conclusion of the novel, which centers on a battle on the airship *Annapurna* commanded by his creator and enslaver, General Caroa. With the help of Nita and her father, who has brokered a parley with Mercier Corporation, Tool infiltrates the flagship for a final battle with the General. Bacigalupi stages this conclusion with multiple nautical resonances. Both the size of the ship – it is one of if not *the* largest airship in the world – and the location (the North Atlantic) – evoke the fate of the *Titanic*, long understood as a metonym for human hubris regarding their technology in the face of nature. Readers familiar with Herman Melville's *Benito Cereno* will also recognize many echoes of the fateful rebellion on the *San Dominick*. Tool's success on the *Annapurna* relies upon humans' refusal to countenance the idea of disobedience among augments, much as Babo and Atufal succeeded in taking the *San Dominick* by relying on whites' underestimation of enslaved persons' intellectual capacity and agency. Tool permits himself to be captured by augments and brought to the command bridge, but then works to awaken them to their own agency. In a manner very different than his winning over the augments of the Patel Corporation, Tool connects to the Mercier augments here through conversation and reasoning, rather than a command:

> We are brothers...
> Are we slaves to do our masters' bidding? Whose wars do we fight?...
> Whose blood is shed?...
> Who will you fight for, brother?[60]

The Mercier augments refuse to harm Tool after contemplating this reasoning for a few tense moments. This success implies that other augments can break the conditioning on their own without the need for an exceptional power to help them override it. The outcome of this liberation proves violent only because the humans turn on the augments when they disobey. Once the humans attack them, the augments consider the oaths broken and retaliate with overwhelming force, sinking the airship by destroying its flotation system and killing every crew member with the exceptions of General Caroa and the ship's chief intelligence analyst, Arial Jones, who face a confrontation with Tool instead. This decimation of the Mercier Corporation's highest ranks provides a dramatic illustration of Dipesh Chakrabarty's prediction that while global elites can stave off the effects of climate change for themselves for a time, eventually they will discover that in a changed Earth, "there are no lifeboats... for the rich and the privileged"[61] In contrast to the humans, the augments are survivors. At the end of their mutiny, Tool instructs the head Mercier augment, Titan, to swim with his kin to Greenland: "Titan would get his kin to safety. They were too strong to do anything but survive. Perhaps they would form an outpost of independence. Take Greenland for themselves. Tool liked the idea, and wished them well."[62] The implication that the Mercier augments will form their own settlement suggests an optimistic vision of a continued free existence for them, with perhaps the suggestion that this place could be the seed for future liberation.

The final confrontation between Tool and General Caroa offers a dramatic revelation of their connection, one that echoes the broader arc of human hubris with respect to nature, embodied in Tool's origin. The general reveals that he has spliced his own DNA into Tool as a way to maintain power over him, with an assertion of quasi-paternal ownership: "You are mine! My blood! My kin! My pack! MINE!"[63] Caroa's claim that Tool is "mine" is a grimly perfect synecdoche for human claims for dominion. Throughout the Drowned Cities trilogy, humans have extracted, exploited, and consumed ecosystems, animals, and other humans because of the belief that the earth belongs to them. Tool is stunned by this revelation and falters in combat, but fights this influence by contemplating how his own bonds throughout the trilogy did not depend upon blood or control: he "had forged different alliances with human beings, with other packs. And those packmates had fought beside him. And protected him. Had risked for him. They had been people. Simple people. Human. Not his kin. Not his blood."[64] Tool's understanding of "pack" offers a model of community based on pledged allegiance, with a recognition that cooperation across species is not only possible, but was necessary for their mutual survival.

In the final moments of the *Annapurna*, Tool opts to preserve the last human survivor, Jones, because of a mutual pledge they make to one another. He will help her swim to safety, while she will find a way to permit him and his crew to escape corporate entanglements once they return to land. Besides securing her own survival, Jones agrees to this plan because she knows it would destroy Mercier's standing among corporations if it were leaked that their augments rebelled against them. The veneer of human mastery over augments is so important to the corporate elite that their leaders would rather paper over a rebellion than actually reckon with the moral complexities that gave rise to it. As Tool swims away from the *Annapurna* with Jones, there is a moment of kindness that echoes his helping Mahlia, but this time it is for the sake of an opponent rather than an ally. Because he has superior lung capacity, Tool tries to share his breath with Jones, and he then resuscitates her on a piece of flotsam with a single, powerful chest compression. Tool's healing breath and this scene of revival may offer a nod to the healing powers of the lion Aslan in C.S. Lewis's *Chronicles of Narnia*. Once she revives, Jones apologizes for her complicity in the murder of augments with the grimly resonant excuse "I followed orders." Tool responds with surprising mercy after reflecting upon the differences and commonalities between humans and augments.

So many human beings, all struggling to survive. So many people doing terrible things, hoping to last another day. A strange thing, this human need for comfort. The human desire to be freed of sin.

You are flawed, he thought.
But to his own surprise, he grasped her hand.
We are flawed.[65]

Bacigalupi stages a powerful twist on expectations of natural selection and extinction in this passage. For all his physical power and combat prowess, Tool rejects claims to mastery or evolutionary superiority over humans at the end. Instead, he finds common cause in the shared fragility and flaws of humans and augments alike. Indeed, he survives at the end not because he can reach land himself, but because Nita and Nailer, whom he had helped in the first novel, have been stealthily sailing underneath the airship in case he needs rescue. Tool's last words to himself in the novel are hopeful: "Humans, working to save him. Kin, if not in blood, then in kind. Pack."[66] Tool's recognition that humans and non-humans can be "kin in kind," offers a rebuttal to the visions of genocide and extinction that have circulated among humans when confronted with the idea of augment rights, and the fact that Tool's final word in the trilogy is "pack" suggests Bacigalupi's vision that only by working with nature and seeing kinship across the lines of species, can humanity survive.

This conclusion reveals that the augment revolution is not complete, but it is a victory that points to a continuing and evolving post-human future, with the outcome likely set by how humans address their relationship to nature. Poor human stewardship has led to a transformed Earth in which humans have fallen from their apex position in the ecosystem and are primarily living via scavenge or subsistence. While there is a small global elite, the fact that they must live on airships or the diminishing high ground suggests that their way of life is unsustainable. In addition, the elite's reliance on exploited augment labor makes their future precarious, as the demonstrated by the sinking of the *Annapurna*, which, like most of the cities in the collection, ends up drowned. As the trilogy concludes, Tool and his crew head out in anonymous ship to secure a new life outside of corporate wars, while Jones resumes her work at Mercier Corporation. This ending echoes the conclusion of *Benito Cereno*, with the critical difference that these mutineers survive. The reader's final glimpse of Tool and his "pack" through Jones's eyes emphasizes the endurance of the eclectic crew:

> On [the] deck, a group of sailors were gathered. An augment loomed amongst them, towering over the smaller human forms... One of the crew seemed to sport mechanical legs, sleek, curving, metallic things.

Another… perhaps it was the light that made it seem that her arm was mechanical as well, black and gleaming in the sun."[67]

Although Jones does not recognize these characters, the reader can discern that Tool has rejoined Mahlia and her friend Ocho, allies who acquired prosthetic limbs across the trilogy after battles and wounds. This tableau embodies Bacigalupi's rejection of an anthropocentric approach to personhood based on biological, mechanical, or ableist norms. Tool, Mahlia, and Ocho are all, in effect, hybrids, and they have not only survived at the end, but seem primed to evade the norms and persecution of the elite in the future. Their diverse alliance, in microcosm, provides an ideal example of post-humanist philosopher Rosi Braidotti's "biocentered egalitarianism," which "entails the displacement of anthropocentrism and the recognition of transspecies solidarity on the basis of our being in this together – environmentally based, embodied, embedded, and in symbiosis."[68] Jones, in contrast, faces a future devoid of solidarity and choked by the fear of imminent violence. She reassures herself that the Mercier Corporation augments around her are loyal, but her last words in the novel suggest she will live with perpetual doubt: "*They won't attack. She could almost convince herself of it.*"[69] The conclusion of the trilogy provides a powerful cautionary tale about the failures of stewardship and the danger of human claims for ownership over the earth. Bacigalupi offers readers a stark choice: seek to dominate, and be doomed; or cooperate, and have a chance at survival. If humans do not respect nature and reject domination and extraction, we will drown not only our cities, but our species, and the earth will abide without us.

Notes

1 Paolo Bacigalupi, *Tool of War* (New York: Little, Brown Books, 2017), 1.
2 "Apocalypse," *Oxford English Dictionary*, accessed June 15, 2021, https://www-oed-com.proxy.ulib.csuohio.edu/view/Entry/9229?redirectedFrom=apocalypse#eid.
3 For example, the Haida (an indigenous nation in the Pacific Northwest) prophesy that if humans disrespect the ocean, the sea will overrun the land and drown most of humanity. In a traditional story of the Papago, from the American Southwest, humans survive a flood, but then challenge the almighty by mining the coral, jet, turquoise, and mother-of-pearl in their region, building a monumental tower. The Great Mystery Power summons the white conquistadores as punishment for their hubris and misuse of the earth's resources. See "The Flood" and "Montezuma and the Great Flood," in *American Indian Myths and Legends*, ed. Richard Erdoes and Alfonso Ortiz (New York: Pantheon Books, 1984), 472–3, 487–9.
4 See Stephen Vincent Benet's "The Place of the Gods," *Saturday Evening Post*, July 31, 1937. The protagonist of this story, a young man named John (whose name echoes the author of Revelation at the end of the Christian New Testament), explores the ruins of a city that was once New York, learning that

the once-great city was destroyed by aerial bombardment and poisonous gases – clear echoes of the weaponry of World War I.
5 There are numerous examples of Cold War novels exploring the aftermath of a nuclear apocalypse. Prime examples include John Wyndham's *The Chrysalids* (London: Michael Joseph, 1955), Nevil Shute's *On the Beach* (New York: William, Morrow and Company, 1957), Pat Frank's *Alas Babylon* (Philadelphia: J.B.Lippincott, 1959), and David Brin's *The Postman* (New York: Bantam Books, 1985).
6 For the archetypal pandemic apocalypses in Cold War American literature, see George R. Stewart's *Earth Abides* (New York: Random House, 1949) and Richard Matheson's *I Am Legend* (New York: Gold Medal Books, 1954). *I Am Legend* is also notable as the originator of an undead apocalypse, as the bacillus spread in World War III turns infected humans into vampires. George Romero later cited the novella as the inspiration for his 1968 film *Night of the Living Dead*, which is widely acknowledged as inaugurating the genre of the zombie apocalypse.
7 See Kyle Bishop, "Dead Man Still Walking: Explaining the Zombie Renaissance," *Journal of Popular Film and Television* 37, no. 1 (Spring 2009): 16–25; Neil McRobert, "'Shoot Everything that Moves:' Post-Millennial Zombie Cinema and the War on Terror," *Textus: English Studies in Italy* 25, no. 3 (September–December 2012): 103–116; Steven Pokornowski, "Insecure Lives: Zombies, Global Health, and the Totalitarianism of Generalization," *Literature and Medicine* 31, no. 2 (Fall 2013): 216–234; Jason Morrisette, "Zombies, International Relations, and the Production of Danger: Critical Security Studies versus the Living Dead," *Studies in Popular Culture* 36, no. 2 (Spring 2014): 1–27; Tamsin Phillips, "Zombies as an Allegory for Terrorism: Understanding the Social Impact of Post 9/11 Security Theatre and the Existential Threat of Terrorism through the Work of Mira Grant," *Law and Literature* 33, no. 1 (2021): 119–140.
8 J.G. Ballard's *The Drowned World* (New York: Berkley Books, 1962) was unusually prescient in its time for focusing on rising sea levels and warming temperatures, rather than nuclear war, as an apocalyptic threat.
9 Early chapters in the first volume of Atwood's trilogy describe the environmental degradation: "as time went on… the coastal aquifers turned salty and the northern permafrost melted and the vast tundra bubbled with methane, and the drought in the midcontinental plains went on and on, and the Asian steppes turned to sand dune." Margaret Atwood, *Oryx and Crake* (New York: Random House, 2003), 24.
10 The collective narrator/chorus of *On Such a Full Sea* advises the reader of the intemperate climate of their world: "Everyone knows it is rough living in the open counties. In this region, where it can get both very hot and very cold, it's especially unpleasant. Though it seems that's most places now! Our elders will say there used to be whole seasons in between of perfectly glorious days. Now, of course, those days are few mere intermittent glimpses of what seems to be us a prehistoric world." See Chang-Rae Lee, *On Such a Full Sea* (New York: Penguin Books, 2014), 13.
11 The first volume of the trilogy lays out the foundational narrative tied to climate change. The mayor of District 12 "tells of the history of Panem, the country that rose up out of the ashes of a place that was once called North America. He lists the disasters, the droughts, the storms, the fires, the encroaching seas that swallowed up so much of the land, the brutal war for what little sustenance remained." Suzanne Collins, *The Hunger Games* (New York: Scholastic, 2008), 18.

12 "Dystopia," *Oxford English Dictionary*, accessed June 17, 2021, https://www-oed-com.proxy.ulib.csuohio.edu/view/Entry/58909?redirectedFrom=dystopia&.
13 Adam One, the founder of the Gardeners, summarizes their mission with a strong echo of the exceptionalism of Puritan settlers from the seventeenth century: "five years ago, this Edencliff Rooftop Garden of ours was a sizzling wasteland, hemmed in by festering city slums and dense of wickedness, but now it has blossomed as the rose. By covering such barren rooftops with greenery, we are doing our small part in the redemption of God's Creation from the decay and sterility around us, and feeding ourselves with unpolluted food in the bargain." Margaret Atwood, *Year of the Flood* (New York: Random House, 2009), 11.
14 Explaining his rationale for developing a pill for mass sterilization of humans (which turns out to be in fact a pill for their mass extinction), Crake tells his friend Jimmy, "As a species we're in deep trouble, worse than anyone's saying. They're afraid to release the stats because people might just give up, but take it from me, we're running out of space-time. Demand for resources has exceeded supply for decades in marginal geopolitical areas, hence the famines and droughts; but very soon, demand is going to exceed supply *for everyone*. With the BlyssPluss Pill the human race will have a better chance of swimming." Atwood, *Oryx and Crake*, 295.
15 Margaret Atwood, *MaddAddam* (New York: Random House, 2013), 358.
16 Deborah C. Bowen, "Ecological Endings and Eschatology: Margaret Atwood's Post-Apocalyptic Fiction," *Christianity and Literature* 66, no. 4 (2017), 697.
17 The National Oceanic and Atmospheric Administration predicts that sea levels could rise by at least one foot, and up to eight feet, by 2100, depending upon whether or not we reduce our emissions that increase global warming. Rebecca Lindsey, "Climate Change: Global Sea Level," accessed August 9, 2021, https://www.climate.gov/news-features/understanding-climate/climate-change-global-sea-level. To put that number into vivid context, an 8-foot sea level rise would drown most of the East Coast, including Boston, Miami, Manhattan, Philadelphia, Norfolk, Charleston, and Miami. See "Sea Levels Could Rise 8 Feet: See How That Will Drown Your City," accessed August 9, 2021, https://www.fastcompany.com/3067846/see-how-the-coming-8-feet-of-sea-level-rise-will-drown-your-city.
18 Paolo Bacigalupi, *Shipbreaker* (New York: Little, Brown Books, 2010), 212.
19 Bacigalupi, *Shipbreaker*, 6.
20 Bacigalupi, *Shipbreaker*, 80.
21 Johan Höglund, "Challenging Ecoprecarity in Paolo Bacigalupi's Ship Breaker Trilogy," *Journal of Postcolonial Writing* 56, no. 4 (2020), 453.
22 Bacigalupi, *Shipbreaker*, 49, 155.
23 I am indebted to Jessica Tresko, an accomplished English teacher and educational advisor with an interest in zoology and ecosystems, for introducing me to the concepts of predator cannibalism and intraspecific predation.
24 "[T]he coywolv had evolved to fill niches that had opened up in a damaged and warming world – all the size and cooperation of a wolf, all the intelligence and adaptability of a coyote. Coywolv had come loping down out of Canada's black winter darkness and then just kept spreading." Paolo Bacigalupi, *Drowned Cities* (New York: Little, Brown Books, 2012), 48.
25 Bacigalupi, *Drowned Cities*, 15.
26 Bacigalupi, *Drowned Cities*, 232.
27 According to Mikhail Bakhtin, "[i]n the literary artistic chronotope, spatial and temporal indicators are fused into one carefully thought-out, concrete whole. Time, as it were, thickens, takes on flesh, becomes artistically visible;

likewise, space becomes charged and responsive to the movements of time, plot, and history." Mikhail Bakhtin, "Forms of Time and of the Chronotope in the Novel," *The Dialogic Imagination*, trans. Caryl Emerson and Michael Holquist (Austin: University of Texas, 1981), 84.
28 Bacigalupi's dialogue between Nita and Tool as they approach Orleans clearly echoes this slave-trading history. In a debate about corporate responsibility, Tool comments with sarcasm about the fact that trading oil is forbidden, but trafficking augments is not: "It's black-market fuel. Banned by convention, if not in fact. The only thing that would be more profitable is shipping half-men, but that of course is legal." Bacigalupi, *Shipbreaker*, 193.
29 Bacigalupi, *Shipbreaker*, 62, 287.
30 Bacigalupi, *Shipbreaker*, 199.
31 Bacigalupi, *Shipbreaker*, 68.
32 Bacigalupi, *Shipbreaker*, 188, 197–198.
33 Nicole Waligora-Davis, *Sanctuary: African Americans and Empire* (New York: Oxford UP, 2011), 139.
34 Bacigalupi, *Shipbreaker*, 204.
35 Bacigalupi, *Shipbreaker*, 232
36 Bacigalupi, *Drowned Cities*, 3, 138.
37 Bacigalupi, *Drowned Cities*, 379.
38 Bacigalupi, *Drowned Cities*, 433.
39 Bacigalupi, *Drowned Cities*, 46.
40 Bacigalupi, *Drowned Cities*, 310.
41 Bacigalupi, *Drowned Cities*, 259, 267–8.
42 Bacigalupi, *Drowned Cities*, 344.
43 Bacigalupi, *Tool of War*, 10.
44 Bacigalupi, *Tool of War*, 15.
45 Bacigalupi, *Tool of War*, 117.
46 Bacigalupi, *Tool of War*, 52.
47 Bacigalupi, *Tool of War*, 275.
48 Bacigalupi, *Tool of War*, 275.
49 Bacigalupi, *Tool of War*, 131.
50 Bacigalupi, *Tool of War*, 172.
51 Bacigalupi, *Tool of War*, 166.
52 Bacigalupi, *Tool of War*, 276, 277, 278.
53 Bacigalupi, *Tool of War*, 287.
54 Bacigalupi, *Tool of War*, 161, 280, 281.
55 Bacigalupi, *Tool of War*, 304.
56 Bacigalupi, *Tool of War*, 303.
57 Bacigalupi, *Tool of War*, 12.
58 Bacigalupi, *Tool of War*, 133.
59 Bacigalupi, *Tool of War*, 209, 221.
60 Bacigalupi, *Tool of War*, 329.
61 Dipesh Chakrabarty, "The Climate of History: Four Theses," *Critical Inquiry* 35, no. 2 (2009): 221.
62 Bacigalupi, *Tool of War*, 341.
63 Bacigalupi, *Tool of War*, 349.
64 Bacigalupi, *Tool of War*, 358.
65 Bacigalupi, *Tool of War*, 363.
66 Bacigalupi, *Tool of War*, 368.
67 Bacigalupi, *Tool of War*, 373.
68 Rosi Braidotti, "Animals, Anomalies, and Inorganic Others," *PMLA* 124, no.2 (Spring 2009): 530.
69 Bacigalupi, *Tool of War*, 372.

Bibliography

Atwood, Margaret. *Oryx and Crake*. New York: Random House, 2003.
Atwood, Margaret. *MaddAddam*. New York: Random House, 2013.
Atwood, Margaret. *Year of the Flood*. New York: Random House, 2009.
Bacigalupi, Paolo. *Drowned Cities*. New York: Little, Brown Books, 2012.
Bacigalupi, Paolo. *Shipbreaker*. New York: Little, Brown Books, 2010.
Bacigalupi, Paolo. *Tool of War*. New York: Little, Brown Books, 2017.
Baktin, Mikhail. "Forms of Time and of the Chronotope in the Novel." *The Dialogic Imagination*. Translated by Caryl Emerson and Michael Holquist. Austin: University of Texas, 1981.
Ballard, J.G. *The Drowned World*. New York: Berkley Books, 1962.
Benet, Stephen Vincent. "The Place of the Gods." *Saturday Evening Post*, July 31, 1937.
Bishop, Kyle. "Dead Man Still Walking: Explaining the Zombie Renaissance." *Journal of Popular Film and Television* 37, no. 1 (Spring 2009): 16–25.
Bowen, Deborah C. "Ecological Endings and Eschatology: Margaret Atwood's Post-Apocalyptic Fiction." *Christianity and Literature* 66, no. 4 (2017): 691–705.
Braidotti, Rosi. "Animals, Anomalies, and Inorganic Others." *PMLA* 124, no. 2 (Spring 2009): 526–532.
Brin, David. *The Postman*. New York: Bantam Books, 1985.
Chakrabarty, Dipesh. "The Climate of History: Four Theses." *Critical Inquiry* 35, no. 2 (2009): 197–222.
Collins, Suzanne. *The Hunger Games*. New York: Scholastic, 2008.
Erdoes, Richard and Alfonso Ortiz, eds. *American Indian Myths and Legends*. New York: Pantheon Books, 1984.
Frank, Pat. *Alas Babylon*. Philadelphia: J.B. Lippincott, 1959.
Höglund, Johan. "Challenging Ecoprecarity in Paolo Bacigalupi's Ship Breaker Trilogy." *Journal of Postcolonial Writing* 56, no. 4 (2020): 447–459.
Lee, Chang-Rae. *On Such a Full Sea*. New York: Penguin Books, 2014.
Lindsey, Rebecca. "Climate Change: Global Sea Level." National Oceanic and Atmospheric Administration. Accessed August 9, 2021. https://www.climate.gov/news-features/understanding-climate/climate-change-global-sea-level.
Lu, Marie. *Legend*. New York: G.P. Putnam's Sons, 2011.
Matheson, Richard. *I Am Legend*. New York: Gold Medal Books, 1954.
McRobert, Neil. "'Shoot Everything that Moves:' Post-Millennial Zombie Cinema and the War on Terror." *Textus: English Studies in Italy* 25, no. 3 (September–December 2012): 103–116.
Morrisette, Jason. "Zombies, International Relations, and the Production of Danger: Critical Security Studies versus the Living Dead." *Studies in Popular Culture* 36, no. 2 (Spring 2014): 1–27.
Oxford English Dictionary. "Apocalypse." Accessed June 15, 2021a. https://www-oed-com.proxy.ulib.csuohio.edu/view/Entry/9229?redirectedFrom=apocalypse#eid.
Oxford English Dictionary. "Dystopia." Accessed June 17, 2021b. https://www-oed-com.proxy.ulib.csuohio.edu/view/Entry/58909?redirectedFrom=dystopia&.

Peters, Adele. "Sea Levels Could Rise 8 Feet: See How That Will Drown Your City." *Fast Company*. Accessed August 9, 2021. https://www.fastcompany.com/3067846/see-how-the-coming-8-feet-of-sea-level-rise-will-drown-your-city.
Phillips, Tamsin. "Zombies as an Allegory for Terrorism: Understanding the Social Impact of Post 9/11 Security Theatre and the Existential Threat of Terrorism through the Work of Mira Grant." *Law and Literature* 33, no. 1 (2021): 119–140.
Pokornowski, Steven. "Insecure Lives: Zombies, Global Health, and the Totalitarianism of Generalization." *Literature and Medicine* 31, no. 2 (Fall 2013): 216–234.
Shute, Nevil. *On the Beach*. New York: William, Morrow and Company, 1957.
Stewart, George R. *Earth Abides*. New York: Random House, 1949.
Waligora-Davis, Nicole. *Sanctuary: African Americans and Empire*. New York: Oxford University Press, 2011.
Wyndham, John. *The Chrysalids*. London: Michael Joseph, 1955.

Part III
Approaches to Contemporary Challenges

7 Political Aspects of Stewardship for Wildlife in the U.S.

Bruce Rocheleau

In 1973, President Nixon signed the Endangered Species Act (ESA) into law "to provide a program for the conservation of endangered and threatened species" because "they have been rendered extinct as a consequence of economic growth and development untempered by adequate concern and conservation"[1] The Act covers virtually all wildlife not just popular mammals and birds but mollusks, crustaceans, and insects – the only exception being insects designated officially as pests. The Act forbids anyone to "harass, harm, pursue, hunt, shoot, wound, kill, trap, capture, or collect" endangered species. The Act enables the government to purchase land and designate "critical habitat" necessary to protect these species and requires that monitoring and "recovery plans" be established for species. The language of the ESA is forceful and does not allow economic costs to be considered in listing species to be protected. In short, this Act codifies a system of stewardship of humans for other species that is impressive in its comprehensiveness and apparent willingness to prioritize the needs of endangered species over human activities that would threaten them. The ESA's significance became even greater because similar laws have been adopted by many other countries throughout the world. This chapter investigates the politics involved with the attempt to provide stewardship over wildlife. It presents an intriguing question: since it is difficult to provide protections for many humans (e.g., minorities) who participate directly in the body politic, it would seem even more challenging to get society to commit to stewarding flora and fauna that have no direct participation or "political resources of their own" and whom humanity has viewed for most of its existence as resources to be exploited. In short, how did such a powerful system of protections for wildlife evolve and how have these stewardship efforts worked in practice?

Origins of Ideas of Protecting Wildlife

The idea of protecting wildlife dates back to early Colonial days of the U.S. – by the time of the Revolutionary War, all colonies except Georgia had closed seasons for deer.[2] By 1850, some states had taken steps to protect non-game birds.[3] But the major impetus for state wildlife management was to protect game animals for wealthier sportsmen and laws were enacted to prevent commercial hunting which was done primarily by blue-collar hunters who were often immigrants.[4] The Boone & Crockett Club, made up of wealthy sportsmen like Teddy Roosevelt, lobbied to protect big game.[5] By the 1860s, organizations began to arise such as the Ornithological Association (1873), the American Society for Prevention of Cruelty to Animals (ASPCA) (1866), and the Audubon Society (1905) that were committed to protecting wildlife for intrinsic reasons and especially birds because threats to their existence had become apparent.[6]

Although some states (e.g., Connecticut & New Jersey) adopted protections for non-game birds, most early wildlife conservation efforts targeted game animals. President Theodore Roosevelt exemplifies somebody who appreciated both the intrinsic value of wildlife and their use for pleasure of trophy hunting. He originated protected monuments and national wildlife refuges such as Pelican Island (Florida) to protect egrets and other birds of interest from plume hunters.[7] At the same time, he also remarked that he wanted to shoot a "free range bison" while they still existed.[8] These alternative value systems coexisted in him and his administration. His Chief of U.S. Forest Service, Gifford Pinchot, was a strong proponent of utilitarianism and viewed wildlife and other natural resources as valuable only insofar as they benefited humanity.[9] Roosevelt also maintained a relationship with John Muir, founder of the Sierra Club, who enunciated a policy of preservationism with appreciation of the intrinsic value of nature apart from its value to humanity. Indeed, Muir criticized Roosevelt for his childish fixation on "killing things."[10] Roosevelt instituted protection for animals in wildlife refuges and established a force to arrest poachers. In addition, he and his fellow Boone & Crockett sportsmen enunciated the principles of "fair chase" that demanded hunters use their skills and abilities to shoot game and not rely on unethical practices such as baiting.

The conflicting nature of these competing principles is illustrated by the Division of Biological Survey (originally in the Department of Interior) that was vested with enforcing the Lacey Act of 1900 aimed at preventing the transportation of illegally killed wildlife. At the same time, however, this agency became dedicated to killing animals such as wolves that preyed upon livestock that grazed in Federal Forest Preserves.[11] By 1915, it took on a more general mission of exterminating wolves and coyotes that affected farmers and ranchers; it is now known as "Wildlife Services" and is located in the Federal Department of Agriculture.[12] The mission of protecting livestock for ranchers won them a strong political constituency that continues till this day.

By the 1930s, scientists and conservationists were emerging to challenge the attack on predators, arguing for the positive biological function performed by them.[13] Aldo Leopold formulated the idea of a "land ethic" in which humanity should work in "harmony" with land and wildlife and not view them simply as a resource to be exploited.[14] However, politically, the utilitarians remained strongly in control as exemplified by President Hoover signing the Animal Damage and Control Act which authorized the extinction of animals that adversely affected agriculture and big game hunting.[15]

Despite the continuing domination of utilitarians and those favoring extinction of predators, scientists continued to do research that established the need to preserve entire ecosystems and not just individual species. Adolph Murie, who worked for the National Park Service and studied both coyotes and wolves, did research that supported the value of these predators. However, the Biological Survey agency refused to accept Murie's research that showed predators did not harm the population of game animals in Alaska, so Murie became adept at politics by supporting the killing of some wolves in McKinley National Park to forestall their overall extinction.[16]

Agricultural and other consummatory interests continued to dominate wildlife policy until the 1960s–1970s when there was an explosion of concern about the environment stimulated by a number of factors, such as threats to iconic species like golden eagles and publication of Rachel Carson's *Silent Spring*, which convinced the public of persistent and growing threats to fauna and flora. Practical politicians began giving attention to environmental interest groups like the Sierra Club and the Audubon Society because their memberships grew enormously during the 1960s and 1970s.[17] Previously, consummatory interest groups, combined with legislative committees and executive agencies, had formed "iron triangles" to control policy made by agencies such as the Bureau of Land Management (BLM) that governed public lands so that ranchers and the energy industry ruled over BLM policy-making.[18] Now this monopoly was challenged by the environmental movement so that a large number of legislative acts were passed during this time period that afforded protection to species and ecosystems, including the Wilderness Act (1964), the National Environmental Policy Act (1969), and the Endangered Species Act (1973).[19] In short, idealistic human commitment to stewardship was ascendant during this period.

Do Wildlife Have Agency?

A major mystery concerning the passage of laws that provided stewardship for wildlife is how fauna and flora could obtain such powerful support for their preservation without any political resources of their own. The solution to this mystery is that humans form strong attachments to wildlife and consequently are willing to spend time and money to lobby on their behalf. The strength of this attachment is remarkable. For example, wild horses have ranged the West for decades and drawn the enmity of powerful ranchers and even some environmentalists because of all of the forage they consume that deprive cattle and other wildlife species of food. Yet they have strong popular support, as evidenced by the passage of the Wild Free-Roaming Horse and Burro Act of 1971 which banned lethal controls of them.[20] The Act declared that the horses were "living symbols of the historic and pioneer spirit of the West; that they contribute to the diversity of life forms within the Nation and enrich the lives of the American people."[21] Although powerful groups such as the Cattlemen's Beef Association pushed to have the wild horse population reduced by whatever means available, including being sold for slaughter, groups such as American Wild Horse Preservation have brought enough pressure to bear to protect them from this fate and instead pushed for the use of birth control to control their numbers. There are many other species that have attracted strong support, with individual groups backing animals such as lions, tigers, whales, apes, pandas, and many birds. Even more obscure species have drawn groups to support them, such as the banana slug of Northern California.[22] Lorimer has provided a detailed case study of how a bird that lives in Scotland, the corncrake, was threatened by development but attracted incredibly devoted followers such as the Royal Society for Preservation of Birds who have brought about changes in farming practices to preserve the bird. In effect, these charismatic animals are able to evoke the formation of human groups that expend great resources to protect them. Thus, in a sense, the animals have "agency."[23]

Of course, not all animals have such positive followings and some animals such as wolves have what I refer to as "negative charisma" – they evoke dislike and hatred that exceeds the amount of harm they do to human interests.[24] Even generally liked animals such as elephants can wreak great damage to the interests of native farmers which results in pressure for their management and culling. So prominent are these conflicts between humans and wildlife that a whole field of study has evolved into "human–wildlife conflict." Catherine Hill argues that the term human–wildlife conflict is a misnomer because the conflict is actually between opposed groups of humans.[25] These human conflicts frequently involve cultural values with wildlife supporters often being outsiders and elites who live far away from the wildlife while rural inhabitants who live near them are opponents of protections. There are national differences in these conflicts – in developing countries, the people most directly affected are often poor natives, whereas in the U.S. wealthy industry groups like energy and livestock industries are most likely to come into conflict with protected species.

One of the limitations of the ESA is its focus on individual species because general public interest in the legislation was largely due to their concern about "charismatic species" such as golden eagles. One of the challenges of supporters of stewardship for wildlife is to convert human attachments from a focus on charismatic species to a commitment to

Political Aspects of Stewardship for Wildlife in the U.S. 139

preserving entire ecosystems, not just a few species. Scientists (e.g., Murie) were already emphasizing the need to focus on overall ecosystems rather than individual species long before the law's passage and research since then has strongly reaffirmed the necessity of this broader approach.[26] The closely interrelated nature of wildlife is illustrated by the phenomena known as "trophic cascade." The disappearance of top predators like wolves led to a decrease in other species such as beavers which then affected many other species lower on the food chain, thus leading to an impoverished biodiversity.[27] Predators are far from being the only problematic wildlife for humans. When sage grouse were being considered to be listed under the ESA, the entire fossil fuel industry in the West was threatened because the bird's range covers millions of acres, much of it in prime areas for drilling and mining that would interfere with the bird's mating.[28] Consequently, the Federal and state governments, along with the energy industry and conservationist groups, worked feverishly to reach a compromise that would prevent the bird's listing while putting restrictions on industry operating near its mating areas (territories known as leks). The contrasting opinions on this compromise are illuminating about perceptions of stewardship. The Audubon Society considered the *avoidance of listing* the sage grouse as the greatest success of the ESA[29] while the environmental group that brought the original suit to protect the bird viewed it as "sellout" and conservatives labeled this agreement an example of "oppressive government."[30] Thus, perceptions of efforts to protect wildlife can vary greatly depending on the groups involved and to what extent compromises are viewed as favoring intrinsic or consummatory interests.

The Politics of the Endangered Species Protection at the U.S. Federal Level

The passage of the 1973 Endangered Species Act occurred due to the fortuitous combination of several simultaneous events. President Nixon was preparing to run for another term and wanted to attract younger voters whom he perceived as environmentalists. Their numbers were to be boosted by a new Constitutional Amendment guaranteeing the vote for anyone 18 years of age. Future opponents of the ESA did not realize its implications, so there was no strong interest-group resistance to its passage. Nixon's senior staff working on the legislation were environmentalists and succeeded in inserting language that eliminated qualifying adjectives such as "practicable" in the Act so that Nixon and conservatives in his party did not realize the significance of the new statute.[31] The legislation passed with virtual unanimous support (92 to 0 in the Senate and 390 to 12 in the House). This consensus quickly broke down when the ESA was used to block the construction of the Tellico Dam in Tennessee due to its threat to an obscure fish (snail darter) and the Supreme Court interpreted the law's protections forcefully. This case was followed by many others in which interests of industries (rancher, developer, energy) were challenged by lawsuits based on ESA protections for wildlife. Thus, the "stewardship consensus" represented by the ESA broke down.

Despite repeated attacks on the ESA by conservative interest groups, including developer and energy industries as well as ranchers, the ESA has survived with its primary clauses intact such as basing decisions about what animals to protect according to the "best science available" and not on their costs to humans. However, there have been significant modifications to the Act which reveal how humans committed to the general goal of stewardship of other species were willing to compromise their idealistic commitment to the intrinsic value of wildlife in order to assure the continuation of the ESA. One major change was to allow developers or other consumptive users to "take" wildlife provided that they participated in a Habitat Conservation Plan (HCP) that "mitigates" the harmful effects of the developments. In the terms of the ESA "take" is defined as "to harass, harm, pursue, hunt, shoot, wound, kill, trap, capture, or collect, or to attempt to engage in any such conduct."[32] HCPs had been created in a 1982 revision of the ESA, but the Reagan and George H.W. Bush administrations had little interest in them. A significant part of Clinton's coalition was the environmental movement, so they created 400 HCPs, compared to only 14 in the previous two Republican administrations.[33]

However, Clinton's administration was acutely aware of the strength of the opposition to the ESA, especially in Western states, and they feared the political consequences of ignoring strong objections of industries to protections that interfered with their livelihoods and profits. Consequently, they adopted other measures to foster compromise between preservationist and consumption interests. Candidate Conservation Agreements (CCAs) and Candidate Conservation Agreements with Assurances (CCAAs) were agreements aimed at areas where there were not formally listed endangered species, but there was a likelihood that threatened species that should be listed as "endangered" did exist. If landowners or other interests agreed to a CCA or CCAAs plan, they would be limited by their future conservation obligations.[34] Still, the big problem remained: how to forge compromises between opposed conservationists and consummatory groups? The Clinton administration adopted a "stakeholder approach" to obtaining compromise. Representatives of both environmentalists and consummatory groups participated in committees that worked together to come to an agreement that was acceptable to both sides. A good case study of such an HCP is provided by Beatley's description of the Coachella Valley California HCP established to preserve the fringe-toed lizard.[35] It reached a successful compromise though many developers remained skeptical of the value of saving the lizard. Success was not guaranteed with this approach, and some failed to reach consensus but a major force for success was fear that failure to participate in the committee would result in failure to have any impact on the final outcome. Clinton referred to this stakeholder approach as a "win–win" solution – it allows both preservationist and consumption sides to claim a "win," though it also implies that both lose too.[36]

However, there is another side to the politics of the ESA: judicial decision-making by the U.S. Federal Courts. While agencies like the U.S. Fish and Wildlife Service (USFWS) employ the compromise-oriented stakeholder approach, the ESA statutory language is strong and allows non-governmental groups to sue to have species protected by formal listing. The ESA law does not allow for compromise based on politics and economics, but decisions are supposed to be made on the basis of the "best available science." So, time and again, environmental groups like the Center for Biodiversity challenged decisions based on scientific criteria and won hundreds of cases, many of them challenging both Republican and Democratic administrations' failure to act to protect species.[37] For example, the Bush, Obama, and Trump administrations have all attempted to delist wolves and grizzlies in certain areas of the West; in the majority of cases, however, these delistings have been invalidated subsequently by the Federal Courts. "Delisting" means that the species is taken off the list of threatened species and thus loses protections. So strong is the wording of the ESA that it has been referred to as a "macho" law,[38] and this is a reason why it has been targeted by conservatives and the Trump administration for revisions that vastly weaken its protections, such as subjecting them to cost–benefit analysis or handing over decisions to state governments. The irony is that the success of the compromise-oriented stakeholder "win–win" approach depends on the existence of the strong ESA law and uncompromising Federal judiciary willing to enforce it if no compromise is reached.[39] Without the existence of this implied threat, industries could afford to stand firm.

In effect, the nationwide consensus that had led to the strong stewardship model of the ESA had broken down with strong industries such as ranchers, farmers, energy producers, and developers unhappy with the law though majorities of the general public and especially the Democratic party were still supportive of it, due their strong environmental constituency. But Clinton was not strongly engaged about the environment and was concerned about the electoral votes of Western and rural states. Bruce Babbitt, Clinton's Secretary of Interior, wrote a book[40] in which he describes his attempt to achieve conservation goals through compromises because he feared that failure to compromise would doom the ESA and that Republican attempts to repeal or weaken it would succeed. Republicans were not strong enough to revise or repeal the ESA law while Democrats could not strengthen it, so a stalemate has occurred that is still not broken. The implementation of the law varied greatly depending on the administration in charge, with Reagan and both Bush administrations being slow to list species as endangered, but even during Clinton and Obama administrations the amount of resources allocated to the primary agency responsible for the Act (U.S. Department of Interior's Fish and Wildlife Service) were small in comparison to the tasks that they were supposed to achieve such as staff for implementing the ESA and dollars to acquire critical habitat. Thus, only a portion of threatened animals actually obtained protection as listed species under the ESA and listings were often only achieved through the action of lawsuits by environmental groups; these took a great deal of time to work their way through courts. Many other agencies are also involved in protecting wildlife but each of them has other interests that are more important to them than stewardship of wildlife such as the Bureau of Land Management (ranching and energy industries) and the Forest Service (logging industry). The amount of resources devoted to restoring species varied greatly, with a few species (often the most charismatic) receiving most of resources devoted to implementation.[41] As an example of the Democratic party's anxiousness to compromise wildlife conservation goals to defuse its angry opponents, one of the first acts of Obama's Secretary of Interior was to delist wolves in Montana and Idaho to appease groups that hated Federal protections for them.[42]

Here is a puzzle: national surveys have consistently shown that a large majority of the U.S. public believes that "protection of the environment should be given priority even at the expense of curbing economic growth" – 65 percent in March of 2019 compared with only 30 percent that favored economic growth.[43] So why have even Democratic administrations been so timid in protecting wildlife? Institutions at the Federal level contribute to the domination of consumptive interests over goals of stewardship when the two conflict. In Congress, senators and representatives have choices as to which committees to serve on and those from Western states flock to those which govern public lands because they are disproportionately located in the West and of prime interest to energy and ranching industries that are major contributors to political campaigns in these states. The Federal agency vested with implementing the ESA (the USFWS) is viewed as a relatively weak agency with its interest group support coming from not-for-profit environmental groups.[44] Its goals regularly bring it into conflict with much more powerful groups with strong profit incentives to oppose its restrictions such as energy and ranching groups. This helps to explain why even under Democratic administrations, support for the ESA has been lukewarm. In summary, at the Federal level, it has been the Federal Courts that have been willing to prioritize wildlife interests over those of strong industries. Both Democratic and Republican administrations and their agencies are willing to compromise their interests in the face of strong opposition.[45]

The Politics of Stewardship at the State Level

The U.S. has a strong system of federalism and, indeed, state agencies play a dominant role in the day-to-day protection of species though the ESA predominates when it comes into conflict with state actions. While popular opinion, together with the existence of the ESA and federal courts, have provided strong support for species when they conflict with powerful economic interests at the federal level, this has been much less the case at the state level. The weakness of stewardship protection at the state level is not due to lack of general popular support. Surveys show that strong majorities in even conservative states favor protection of species over economic interests. The contradiction between overall popular support and actual policy is illustrated by Idaho's experience with wolves. A 1992 Idaho survey found over 70 percent of residents supported the reintroduction of wolves into the state, but the Idaho State Legislature forbade its Fish and Game Department from working with the Federal government to achieve this. Furthermore, a long-time governor of Idaho (2007–2019), Butch Otter, built a successful political career out of opposing wolves by, for example, refusing to prosecute poachers of them.[46] Another example is Wisconsin. Surveys of the state's public show that the majority favored expansion of wolves but when the state wildlife management agency held meetings on wolf policy, wolf opponents outnumbered supporters by a 6:1 margin.[47] There is a general principle that helps to explain such outcomes: when the costs of a policy are concentrated on the consumptive interests of a small group of individuals, they become much more highly engaged than the larger masses who support intrinsic and idealistic values, albeit with much less intensity. In short, this majority is less likely to turn out at such meetings and less likely to affect their votes for candidates than the smaller groups that oppose such conservation because it harms their economic interests.[48] Consequently, opponents of protections and conservation efforts at the state level often prevail.

Although majorities of the public favor wildlife conservation, there are significant differences between urban and rural areas. For example, Colorado held a ballot initiative in 2020 over whether to reintroduce wolves to the state. Preliminary surveys showed that a large majority in the State (84 percent) favored the initiative; when the vote took place, however, it passed by only a slim margin (50 to 49 percent).[49] Urban areas voted strongly in favor but rural areas (near where the wolves would mainly reside) voted against it. Major donors to the opposition campaign included the Colorado Farm Bureau, the Rocky Mountain Elk Foundation, and Safari Club International. The geographical divide in these votes supports the importance of identification with place that Tyra Olstad describes in another essay in this volume.[50]

There are other important factors that explain these outcomes: institutional and funding arrangements at the state levels favor consummatory interests. State wildlife management departments are often strongly dominated by hunters. Indeed, membership wildlife policy advisory groups often specify that a large portion of their members must be hunters or farmers.[51] For example, Wisconsin's Natural Resources Board has the following requirements:

> Wisconsin Act 149 states that beginning May 1, 2017, at least 1 Board member must have an agricultural background and at least 3 Board members must have held a hunting, fishing, or trapping license in at least 7 of the 10 years before the year of nomination except if an individual served on active duty in the U.S. armed forces or national guard.[52]

Thus, groups advising state wildlife departments are often dominated by consummatory group members. Such institutional arrangements generally outweigh the impact of general popular opinion.

Likewise, the funding system of state wildlife management is also crucial to the dominance of consummatory groups. The majority of their funds come from state hunting and fishing licenses or grants from the Federal government (the Pitman–Robertson Act) based on taxes on hunting and fishing equipment. Since most state wildlife management agency funds come either directly through hunting-fishing licenses or indirectly through taxes on firearms and ammunition, these groups feel entitled to dominate policy-making and they have consistently done so. For decades, U.S. sportsmen have boasted about the effectiveness of this funding arrangement that supports hunting-fishing on public lands – they refer to it as the North American Model of wildlife conservation that opposes market-based private commercial activity.[53]

However, there is strong evidence that the basis of this funding is failing due to a long-term (since the 1960s till present) decline in both the percentage and numbers of hunters in the U.S. The size of the decline is huge. For example, in conservative Utah, the number of hunting licenses dropped from 114,000 in 1986 to 16,000 in 2013.[54] This decline occurred despite the fact that Utah's Division of Wildlife Resources, like most other state wildlife agencies, is making great efforts to retain and recruit hunters such as giving partridge chicks to families so their children can participate in their first hunts. The big decline in license and excise taxes on guns and ammunition has necessitated cutbacks in the budgets of many state wildlife agencies and some have added small general taxes to replace a portion of these diminished funds.[55] However, many hunters have resisted adding new general (i.e., not related to hunting) revenue sources because they feared that it would result in their loss of control over policy such as in Montana where a proposed stamp whose revenues would be devoted to activities preventing wolf–human conflicts was rejected by the hunting community for this very reason.[56]

Thus, state wildlife management remains generally an "inside affair" dominated by key interest groups such as hunters, anglers, farmers, and ranchers. There are some cases when key wildlife policies are subjected to public vote. Some states allow groups to put "initiatives" on state-wide ballots over controversial wildlife issues such as trapping, baiting, use of dogs, or hunting of specific species such as mourning doves.[57] A coalition of hunting interest groups (e.g., the National Rifle Association, the Safari Club, and the Rocky Mountain Elk Foundation) have funded opposition to such initiatives while animal welfare organizations like the Humane Society have supported them. One study found a high rate of success of restrictions on hunting during the 1990s.[58] Analysis by Duda et al. found that conservationists were most successful when the practice at issue conflicted with the "fair chase" ethic valued by Teddy Roosevelt.[59] In short, when protections for wildlife make it to the ballot box, the majority views, as measured by poll results, have a good chance of winning, but only eleven states allow such initiatives and thus most decisions about wildlife are made in agency offices, not through voting.

Ranchers are the other group that has been most active in exerting control over wildlife policy in Western states. As with hunters, their numbers are dwindling and their importance to the economy of Western states has declined. Indeed, the West is now nearly as urban as the East with about three-quarters of the population living in urban areas. Despite this decline, rancher influence continues to flourish and dominate in many states due to institutional factors. Like hunters, they are a heavy presence on state wildlife agency boards and also on the advisory committees of the Bureau of Land Management that is in charge of millions of acres. Another significant factor is that many ranchers run for and win state-wide or national office (e.g., Senator Paul Laxalt (Arizona), Senator Alan Simpson (Wyoming), and Senator Larry Craig (Idaho)) and they actively support the interests of ranchers while in office.

The Current State of Stewardship in the U.S.: The Trump and Biden administrations

Many people expected that a "new West" would emerge with a strong conservationist ethic due to its changing demographics and diminishing economic importance of ranching and hunting, but that has not occurred. Indeed, under the Trump administration, consummatory values became ascendant. There were no conservationists in top positions of Trump's Department of Interior which has the major responsibility for the ESA and stewardship of wildlife. For example, Texas rancher Susan Combs who led the fight against the Endangered Species Act and the U.S. Fish & Wildlife Service (USFWS)'s attempt to implement it in Texas, became the Assistant Secretary of Policy, Management and Budget of the USFWS.[60] Likewise, Karen Budd Falen was named Acting Assistant Secretary for Fish, Wildlife and Parks in the Interior Department. She had made a career of advocating for a weakened ESA and has defended public lands ranchers like Cliven Bundy who fought against protection of desert tortoises.[61] In short, under Trump, the Federal agency responsible for protecting wildlife and implementing the ESA was controlled by opponents of the law.[62] The Trump administration gave total commitment to "energy dominance," which meant it favored leasing lands and oceans for drilling and mining with weakened protections for environmental concerns such as species protections. For example, they cut protections against incidental damage for migratory birds under the International Bird Migratory Act and they rejected the concept of requiring compensatory mitigation by industries for actions that harm wildlife which had been introduced during the Obama administration. In short, the Trump administration opposed any conservation or stewardship policies and actions that would endanger profits for industry.[63]

In its early months the Biden administration has reversed several of the Trump policies, such as reinstating protections for migratory birds under the Migratory Bird Treaty. It has also paused leasing public lands for oil and gas exploration and reversed the actions to drill the Artic National Wildlife Reserve for oil. Most significantly, the Biden administration has proposed a program of "30 by 30," meaning that it is committed to protecting at least 30 percent of land and waters for nature by 2030. The concept for this goal originates from biologist Edward Wilson's book, *Half-Earth: Our Planet's Fight for Life*, which argues that humans "are ravenous consumers of all the planet's inadequate bounty" and that they need "a major shift in moral reasoning with a greater commitment given to the rest of life."[64] This idea of giving priority to preserving lands and waters for wildlife and nature is a reversal of Trump's devotion to industries and human profits. However, progress toward such a goal will be difficult. Early issues are already arising such as to what constitutes "conserved" land – for example, does a "working cow ranch" count or should we limit the term to land that is "protected" from such uses?[65] The warming climate is aggravating the crisis for many species. Again, the Biden administration is proposing to address climate, unlike the Trump administration which banned mention of climate from some Interior websites. The success of the Biden administration may depend on the rulings made by the Supreme Court and other federal appeals courts. The Trump administration, in conjunction with Mitch McConnell and the Republican-controlled Senate, appointed a large number of Federal judges as well as three Supreme Court justices. Since the federal courts have been the major bulwark against backsliding on the ESA, the ascendance of Trump appointees could undermine the ESA and other conservation laws.

However, the fact that issues such as "30 by 30" are being seriously considered today reveals how sharp a turn has been taken by the Biden administration from Trump policies. How effective have the ESA and other stewardship protections for wildlife been? Opponents of the law cite the fact that only a small proportion of species that are put on the endangered list subsequently are delisted because they are no longer threatened.[66] These opponents view the relatively small number of delistings as proof of ESA's failure. Defenders of the law point out that threats to these species, such as habitat declines due to development, agriculture, and population growth, are steadily increasing and thus we can expect that many species will remain threatened forever.[67] The USFWS emphasizes that 98 percent of species that have been listed as endangered or threatened continue to exist despite these continually increasing threats and thus the ESA should be considered a success.[68] There have been some spectacularly successful delistings, such as the American alligator, though in this case its success was due to the fact that causes of their decline, poaching, were easier to control than broader underlying causes implicated in the decline of many other species, such as development and agriculture.[69]

If we view stewardship as a scale that can tip toward either preservationist or consumption values, the balance point has varied greatly, depending on the presidential administration in charge. The general populace has strong feelings in favor of the stewardship of wildlife and is willing to sacrifice economic benefits to help wildlife. However, when stewardship threatens the interests of specific individuals with significant costs, these aspirational goals of the majority are often overwhelmed by the political power of consummatory groups – the salience of preservationist is often not high enough to prevail. Still, despite this natural human tendency of people to protect their self-interests, the achievements of the ESA and other conservationist laws are impressive. These protections have persisted despite strong organized opposition and institutional biases that favor selfish interests over idealism. The ESA has helped to save and restore many species by reaching compromises, albeit with the threat of Federal listing of the species due to the willingness of U.S. Federal courts to enforce the law as written according to the best science, and without employing cost–benefit analysis. The 1960s–70s era of ascendancy of concern for stewardship of wildlife occurred in part due to notable threats to iconic species. Unfortunately, the next increase in the salience of concern for wildlife may arise with more extinction threats, as revealed by Cornell University's study that the U.S. bird population has decreased by 29 percent over the past 50 years.[70] As human populations continue to expand and wildlife decline, U.S. citizens may yet come to place higher priority on other species.

Notes

1. *Endangered Species Act of 1973.* (Pub.L. 93–205). https://www.fws.gov/international/pdf/esa.pdf.
2. Peter Matthiessen, *Wildlife in America* (New York: Viking Penguin, Inc., 1987), 65.
3. Matthiessen, *Wildlife in America*, 56.
4. Joe E. Dizard, *Going Wild: Hunting, Animal Rights, and the Contested Meaning of Nature* (Amherst: University of Massachusetts Press, 1999), 111–116.
5. Douglas Brinkley, *The Wilderness Warrior: Theodore Roosevelt and the Crusade for America* (New York: HarperCollins Publishers, 2009), 206.
6. Matthiessen, *Wildlife in America*, 158–165.
7. Brinkley, *Wilderness Warrior*, 16.
8. Brinkley, *Wilderness Warrior*, 150.
9. Brinkley, *Wilderness Warrior*, 579.
10. Brinkley, *Wilderness Warrior*, 544.
11. Michael J. Robinson, *Predatory Bureaucracy: The Extermination of Wolves and the Transformation of the West* (Boulder: The University Press of Colorado, 2005), 61–63.
12. Robinson, *Predatory Bureaucracy*, 103.
13. Curt Meine, "Early Wolf Research and Conservation in the Great Lakes Region," in *Recovering of Gray Wolves in the Great Lakes Region of the United States*, eds. Adrian Wydeven, Timothy R. Van Deelen, and Edward J. Heske (New York: Springer, 2009), 3.
14. Ben A. Minteer, "Valuing Nature," in *Loss of Biodiversity*, ed. Sharon L Spray. & Karen L. McGothlin, (Lanham: Rowman & Littlefield Publishers, Inc. 2003), 79.
15. Robinson, *Predatory Bureaucracy*, 41.
16. Timothy Rawson, *Changing Tracks: Predators and Politics in Mt. McKinley National Park* (Fairbanks: University of Alaska Press, 2001), 41, 142, & 197.
17. Michael E. Kraft, "U.S. Environmental Policy and Politics: From the 1960s to the 1990s," in *Environmental Politics and Policy 1960s–1990s*, ed. Otis L. Graham (University Park: The Pennsylvania State University, 2000), 23.
18. George Hoberg, "The Emerging Triumph of Ecosystem Management: The Transformation of Federal Forest Policy," in *Western Public Lands and Environmental Politics*, ed. Charles Davis (Boulder: Westview, 2001), 58.
19. Kraft, "U.S. Environmental Politics and Policy," 21–23.
20. U.S. General Accountability Office, *Bureau of Land Management: Effective Long-Term Options Needed to Manage Unadoptable Wild Horses* (Washington, DC: USGAO, GAO-09-77, 2008), 1.
21. U.S. General Accountability Office, *Effective Long-Term Options Needed to Manage Unadoptable Wild Horses*, 1–2.
22. James M. Jasper and Dorothy Nelkin, *The Animal Rights Crusade: The Growth of a Moral Protest* (New York: The Free Press, 1992), 51.
23. Jamie Lorimer, *Wildlife in the Anthropocene: Conservation after Nature* (Minneapolis: University of Minnesota Press, 2015).
24. Bruce Rocheleau, *Wildlife Politics* (Cambridge: Cambridge University Press, 2017), 98.
25. Catherine M. Hill, "Introduction: Complex Problems: Using a Biosocial Approach to Understanding Human–Wildlife Interactions," in *Understanding Conflicts about Wildlife*, eds. Catherine M. Hill, Amanda D. Webber, & Nancy E.C. Priston (New York: Berghahn, 2017), 3.

26 National Research Council, *Science and the Endangered Species Act*, (Washington, DC: National Academy Press. 1995), 46.
27 Kristin N. Marshall, N. Thompson Hobbs and David Cooper, "Stream Hydrology Limits Recovery of Riparian Ecosystems after Wolf Reintroduction," *Proceedings of the Royal Society, Biological Sciences*, 280, no. 1756 (April 7, 2013), https://rspb.royalsocietypublishing.org/content/280/1756/20122977.
28 Diane Cardwell and Clifford Krauss, "Frack Quietly Please: Sage Grouse Is Nesting," *The New York Times*, July 19, 2014.
29 Hillary Rosner, "Rethinking the List," *Audubon Magazine*, November-December, 2014, 10.
30 Matthew Brown and Mead Gruver, "Sage grouse plan aims for balance between industry, wildlife," *Associated Press*, September 22, 2015, https://www.seattletimes.com/nation-world/u-s-rejects-protections-for-greater-sage-grouse-across-west/.
31 Joe Roman, *Listed: Dispatches from America's Endangered Species Act* (Cambridge, MA: Harvard University Press, 2011), 52.
32 National Oceanic and Atmospheric Administration, U.S. Department of Commerce, "*Endangered Species Act*," accessed 2 September 2021, https://www.fisheries.noaa.gov/national/endangered-species-conservation/endangered-species-act#section-4-determination-of-endangered-species-and-threatened-species.
33 David J. Sousa and Christopher McGrory Klyza, "New Directions in Environmental Policy Making: An Emerging Collaborative Regime or Reinventing Interest Group Liberalism?" *Natural Resource Journal* 47 (Spring 2007): 385.
34 Martha F. Phelps, "Candidate Conservation Agreements Under the Endangered Species Act: Prospects and Perils of an Administrative Experiment," *Boston College Environmental Affairs Law Review* 25, no. 1 (1997): 175–212.
35 Timothy Beatley, *Habitat Conservation Planning: Endangered Species and Urban Growth* (Austin: University of Texas Press, 1994).
36 Sousa and Klyza, "New Directions in Environmental Policy Making," 426.
37 Rocheleau, *Wildlife Politics*, 151–156.
38 Holly Doremus and A. Dan Tarlock, *Water War in the Klamath Basin: Macho Law, Combat Biology, and Dirty Politics* (Washington: Island Press, 2008).
39 Robert B. Keiter, *Keeping Faith with Nature: Ecosystems, Democracy, & America's Public Lands* (New Haven: Yale University Press, 2003), 118–119; and Martin Nie, *The Governance of Western Public Lands: Mapping Its Present and Future* (Lawrence: University Press of Kansas, 2008), 73.
40 Bruce Babbitt, *Cities in the Wilderness: A New Vision of Land Use in America* (Washington: Island Press/Shearwater Books, 2005).
41 United States General Accounting Office, *Endangered Species: Management Improvement Could Enhance Recovery Program*, Report to the Chairman, Subcommittee on Fisheries and Wildlife Conservation, and the Environment, Committee on Merchant Marine and Fisheries, House of Representatives (Washington, DC: GAO/RCED-89-5, December 1988), 32–33.
42 Jim Robbins, "Target Green: Federal Land Managers Under Attack," *Audubon Magazine*, July–August, 1995, 82–85.
43 Edward O. Wilson, "On *Silent Spring*," in *Courage for the Earth: Writers, Scientists, and Activists Celebrate the Life and Writing of Rachel Carson*, ed. Peter Matthiessen (Boston: Houghton Mifflin Company, 2007), 27–36.

44 Craig W. Thomas, *Bureaucratic Landscapes: Interagency Cooperation and the Preservation of Biodiversity* (Cambridge, MA: Massachusetts Institute of Technology Press), 2003.
45 Keiter, *Keeping Faith with Nature*, 118–119.
46 John Miller, "Idaho Governor Pulls State Management of Wolves in Dispute over Wolf Hunt," *The Associated Press*, October 19, 2010, https://www.flatheadbeacon.com/articles/article/idaho_wont_manage_wolves_under_esa/20166/.
47 James Janega, "Vote to Consider Wolf Hunting Triggers Debate," *Chicago Tribune*, April 27, 2008, https://www.chicagotribune.com/news/ct-xpm-2008-04-27-0804260192-story.html.
48 William Amos, Kathryn Harrison, and George Hober, "In Search of a Minimum Winning Coalition: The Politics of Species-at-Risk Legislation in Canada," in *Politics of the Wild: Canada and Endangered Species*, ed. Karen Beazley and Robert Boardman (Oxford: Oxford University Press, 2001), 137–166.
49 Ballotpedia, "*Colorado Proposition 114, Gray Wolf Reintroduction Initiative*," 2020, https://ballotpedia.org/Colorado_Proposition_114,_Gray_Wolf_Reintroduction_Initiative_(2020).
50 See Tyra Olstad's "Stewardship and Sense of Place: Assumptions and Idealism," above 13–28.
51 Bruce Rocheleau, "The Politics of State Wildlife Management: Why Anti-Conservation Forces Usually Win," accessed 2 September 2021, https://www.wildlifepolitics.org/the-politics-of-state-wildlife-management-why-anti-conservation-forces-usually-winpapers-on-wildlife-politics--policy.html.
52 Wisconsin Department of Natural Resources, Board Member Information, accessed 2 September 2021, https://dnr.wi.gov/about/nrb/members.html.
53 Michele Beucler and Gregg Servheen, "The North American Model of Wildlife Conservation: Affirming the Role, Strength and Relevance of Hunting in the 21st Century: Mirror, Mirror, on the Wall: Reflections from a Nonhunter," in *Transactions of the 73rd North American Wildlife and Natural Resources Conference* (Washington, DC: Wildlife Management Institute, 2009), 163–179.
54 Jennifer Dobner, "Fostering Pheasants to Keep a Tradition Alive," *The New York Times*, November 5, 2014.
55 Rocheleau, *The Politics of State Wildlife Management*.
56 News Bulletin "Lessons from the Montana Wolf Management Stamp," *Wildlife Management Institute* 68, no. 10 (October 2014) https://wildlife-management.institute/outdoor-news-bulletin/october-2014/lessons-montana-wolf-management-stamp.
57 Mark Damian Duda, Martin F. Jones, and Andrea Criscione, *The Sportsman's Voice: Hunting and Fishing in America* (State College, PA: Venture Publishing Inc., 2010).
58 Wayne Pacelle, "Forging a New Wildlife Management Paradigm: Integrating Animal Protection Values," *Human Dimensions of Wildlife: An International Journal* 3, no. 2, (1998): 47.
59 Duda, *The Sportsman's Voice*.
60 Kellie Lunney, Senate confirms Susan Combs in bipartisan vote, *E&E News*, June 5, 2019, https://www.eenews.net/eenewspm/2019/06/05/stories/1060491353.
61 Scott Streater and Jennifer Yachnin, "Trump BLM channels Reagan's Sagebrush Rebellion," *EE News*, August 6, 2019. https://www.eenews.net/greenwire/2019/08/06/stories/1060866919.

62 Kirk F. Siegler, "Critic of Federal Public Lands Management to Join Department of the Interior," *NPR.com*, October 15, 2018, https://www.npr.org/2018/10/15/657542759/critic-of-federal-public-lands-management-to-join-department-of-the-interior.
63 Bruce Rocheleau, *Industry First: The Attack on Conservation by Trump's Interior Department*, Independently published, 2021. https://www.amazon.com/dp/B08TYY51KR?ref_=pe_3052080_397514860.
64 Edward O. Wilson, *Half-Earth: Our Planet's Fight for Life* (New York: Liveright Publishing Corporation, 2016).
65 Jennifer Yachnin, "Does Biden's '30x30' Plan Trade Science for Popularity?," *E&E News*, June 2, 2021, https://www.eenews.net/greenwire/2021/06/02/stories/1063734011?utm_campaign=edition&utm_medium=email&utm_source=eenews%3Agreenwire.
66 U.S. Senate, *Listing and Delisting Processes Under the Endangered Species Act*, Sect. Hearing before the Subcommittee on Fisheries, Wildlife, and Water of the Committee on the Environment and Public Works, 137th Congress First Session on the Regulations and Procedures of the U.S. Fish and Wildlife Service Concerning the Listing and Delisting of Species Under the Endangered Species Act, S. HRG. 107–322 May 9, 2001, www.gpo.gov/fdsys/pkg/CHRG-107shrg78073/html/CHRG-107shrg78073.htm.
67 Wilson, "On *Silent Spring*," 34.
68 U.S. Fish and Wildlife Service, *Endangered Species. Defining Success Under the Endangered Species Act* (Spring 2013), accessed 2 September 2021, https://www.fws.gov/endangered/news/episodes/bu-04-2013/coverstory/index.html.
69 Steven R. Beissinger and John D. Perrine, "Extinction, Recovery, and the ESA," in *Protecting Endangered Species in the United States*, eds. Jason F, Shogren, and John Tschirhart (Cambridge: Cambridge University Press, 2001), 51–71.
70 Kenneth Rosenberg et al., "Decline of the North American avifauna," *Science* 366 (October 4, 2019): 120–124.

Bibliography

Amos, William, Kathryn Harrison, and George Hober. "In Search of a Minimum Winning Coalition: The Politics of Species-at-Risk Legislation in Canada." In *Politics of the Wild: Canada and Endangered Species*, edited by Karen Beazley and Robert Boardman, 137–166. Oxford: Oxford University Press, 2001.

Babbitt, Bruce. *Cities in the Wilderness: A New Vision of Land Use in America*. Washington: Island Press/Shearwater Books, 2005.

Ballotpedia. "Colorado Proposition 114, Gray Wolf Reintroduction Initiative." 2020. https://ballotpedia.org/Colorado_Proposition_114,_Gray_Wolf_Reintroduction_Initiative_(2020).

Beatley, Timothy. *Habitat Conservation Planning: Endangered Species and Urban Growth*. Austin: University of Texas Press, 1994.

Beissinger, Steven R. and John D. Perrine. "Extinction, recovery, and the ESA." In *Protecting Endangered Species in the United States*, edited by Jason Shogren, and John F. Tschirhart, 51–71. Cambridge: Cambridge University Press, 2001.

Beucler, Michele, and Gregg Servheen. "The North American Model of Wildlife Conservation: Affirming the Role, Strength and Relevance of Hunting in the 21st Century: Mirror, Mirror, on the Wall: Reflections from a Nonhunter."

In *Transactions of the 73rd North American Wildlife and Natural Resources Conference*, 163–179. Washington, DC: Wildlife Management Institute, 2009.

Brinkley, Douglas. *The Wilderness Warrior: Theodore Roosevelt and the Crusade for America*. New York: HarperCollins Publishers, 2009.

Brown, Matthew and Mead Gruver. "Sage grouse plan aims for balance between industry, wildlife." *Associated Press*, September 22, 2015. https://www.seattletimes.com/nation-world/u-s-rejects-protections-for-greater-sage-grouse-across-west/.

Cardwell, Diane and Clifford Krauss. "Frack Quietly, Please: Sage Grouse Is Nesting." *The New York Times*, July 19, 2015. https://www.nytimes.com/2014/07/20/business/energy-environment/disparate-interests-unite-to-protect-greater-sage-grouse.html?emc=edit_th_20140720&nl=todaysheadlines&nlid=10365419.

Dizard, Joe E. *Going Wild: Hunting, Animal Rights, and the Contested Meaning of Nature*. Amherst: University of Massachusetts Press, 1999.

Dobner, Jennifer. "Fostering Pheasants to Keep a Tradition Alive." *The New York Times*, November 5, 2014. https://www.nytimes.com/2014/11/06/us/fostering-pheasants-to-keep-a-tradition-alive.html?emc=edit_th_20141106&nl=todaysheadlines&nlid=10365419&_r=0.

Doremus, Holly and A. Dan Tarlock. *Water War in the Klamath Basin: Macho Law, Combat Biology, and Dirty Politics*. Washington: Island Press, 2008.

Duda, Mark Damian; Martin F. Jones, and Andrea Criscione. *The Sportsman's Voice: Hunting and Fishing in America*. State College, PA: Venture Publishing Inc., 2010.

Endangered Species Act of 1973. (Pub.L. 93–205), 1973. https://www.fws.gov/international/pdf/esa.pdf.

Hill, Catherine M. "Introduction: Complex Problems: Using a Biosocial Approach to Understanding Human-Wildlife Interactions." In *Understanding Conflicts about Wildlife*, edited by Catherine M. Hill, Amanda D. Webber, & Nancy E.C. Priston, 1–14. New York: Berghahn, 2017.

Hoberg, George. "The Emerging Triumph of Ecosystem Management: The Transformation of Federal Forest Policy." In *Western Public Lands and Environmental Politics*, edited by Charles Davis, 55–85. Boulder: Westview, 2001.

Janega, James. "Vote to Consider Wolf Hunting Triggers Debate." *Chicago Tribune*, April 27, 2008. https://www.chicagotribune.com/news/ct-xpm-2008-04-27-0804260192-story.html.

Jasper, James M. and Dorothy Nelkin. *The Animal Rights Crusade: The Growth of a Moral Protest*. New York: The Free Press, 1992.

Keiter, Robert B. *Keeping Faith with Nature: Ecosystems, Democracy, & America's Public Lands*. New Haven: Yale University Press, 2003.

Kraft, Michael E. "U.S. Environmental Policy and Politics: From the 1960s to the 1990s." In *Environmental Politics and Policy 1960s-1990s*, edited by Otis L. Graham, 17–42. University Park: The Pennsylvania State University Press, 2000.

Lorimer, Jamie. *Wildlife in the Anthropocene: Conservation after Nature*. Minneapolis: University of Minnesota Press, 2015.

Lunney, Kellie. "Senate Confirms Susan Combs in Bipartisan Vote." E&E News, June 5, 2019. https://www.eenews.net/eenewspm/2019/06/05/stories/1060491353.

Marshall, Kristin N., N. Thompson Hobbs and David J. Cooper. "Stream Hydrology Limits Recovery of Riparian Ecosystems after Wolf Reintroduction." Proceedings of the Royal Society, Biological Sciences 280, no. 1756 (April 7, 2013). http://rspb.royalsocietypublishing.org/content/280/1756/20122977.

Matthiessen, Peter. Wildlife in America. New York: Viking Penguin, Inc., 1987.

Meine, Curt. "Early Wolf Research and Conservation in the Great Lakes Region." In Recovering of Gray Wolves in the Great Lakes Region of the United States. Edited by Adrian Wydeven, Timothy R. Van Deelen and Edward J. Heske, 1–13. New York: Springer, 2009.

Miller, John. "Idaho Governor Pulls State Management of Wolves in Dispute Over Wolf Hunt." The Associated Press, October 19, 2010. https://www.flatheadbeacon.com/articles/article/idaho_wont_manage_wolves_under_esa/20166/.

Minteer, Ben A. "Valuing Nature." In Loss of Biodiversity. Edited by Sharon L. Spray and Karen L. McGothlin, 75–98. Lanham: Rowman & Littlefield Publishers, Inc. 2003.

National Oceanic and Atmospheric Administration. U.S. Department of Commerce. "Endangered Species Act." https://www.fisheries.noaa.gov/national/endangered-species-conservation/endangered-species-act#section-4-determination-of-endangered-species-and-threatened-species.

National Research Council. Science and the Endangered Species Act. Washington, DC: National Academy Press, 1995.

Nie, Martin. The Governance of Western Public Lands: Mapping Its Present and Future. Lawrence: University Press of Kansas, 2008.

Noss, Reed F., Michael A. O'Connell and Dennis D. Murphy. The Science of Conservation Planning: Habitat Conservation Under the Endangered Species Act. Washington, DC: Island Press, 1997.

Otter, C.L. Butch, Governor of Idaho. Letter to Ken Salazar, Secretary of the Interior. 2010 [cited April 16, 2015]; https://fishandgame.idaho.gov/public/docs/wolves/letterGovernor1.pdf.

Pacelle, Wayne. "Forging a New Wildlife Management Paradigm: Integrating Animal Protection Values." Human Dimensions of Wildlife: An International Journal 3, no. 2 (1998): 42–50.

Phelps, Martha F. "Candidate Conservation Agreements Under the Endangered Species Act: Prospects and Perils of an Administrative Experiment." Boston College Environmental Affairs Law Review 25, no. 1 (1997): 175–212.

Rawson, Timothy. Changing Tracks: Predators and Politics in Mt. McKinley National Park. Fairbanks: University of Alaska Press, 2001.

Robbins, Jim. "Target Green: Federal Land Managers Under Attack." Audubon Magazine. July–August, 1995.

Robinson, Michael J. Predatory Bureaucracy: The Extermination of Wolves and the Transformation of the West. Boulder: The University Press of Colorado, 2005.

Rocheleau, Bruce. Wildlife Politics. Cambridge: Cambridge University Press, 2017.

Rocheleau, Bruce. *The Politics of State Wildlife Management: Why Anti-Conservation Forces Usually Win*, 2019. https://www.wildlifepolitics.org/the-politics-of-state-wildlife-management-why-anti-conservation-forces-usually-winpapers-on-wildlife-politics--policy.html.

Rocheleau, Bruce. *Industry First: The Attack on Conservation by Trump's Interior Department*. Independently published, 2021. https://www.amazon.com/dp/B08TYY51KR?ref_=pe_3052080_397514860.

Roman, Joe. *Listed: Dispatches from America's Endangered Species Act*. Cambridge, MA: Harvard University Press, 2011.

Rosenberg, Kenneth et al. "Decline of the North American avifauna." *Science* 366, no. 4 (October 4, 2019): 120–124.

Rosner, Hillary. "Rethinking the List." *Audubon Magazine*, November–December, 2014.

Siegler, Kirk F. "Critic of Federal Public Lands Management to Join Department of the Interior." *NPR.com*, October 15, 2018. https://www.npr.org/2018/10/15/657542759/critic-of-federal-public-lands-management-to-join-department-of-the-interior.

Sousa, David J. and Christopher McGrory Klyza. "New Directions in Environmental Policy Making: An Emerging Collaborative Regime or Reinventing Interest Group Liberalism?" *Natural Resource Journal* 47 (Spring 2007): 378–444.

Streater, Scott and Jennifer Yachnin. "Trump BLM channels Reagan's Sagebrush Rebellion." *EE News*, August 6, 2019. https://www.eenews.net/greenwire/2019/08/06/stories/1060866919.

Thomas, Craig W. *Bureaucratic Landscapes: Interagency Cooperation and the Preservation of Biodiversity*. Cambridge, MA: Massachusetts Institute of Technology Press, 2003.

Tobin, Richard J. *The Expendable Future: U.S. Politics and the Protection of Biological Diversity*. Durham: Duke University Press, 1990.

U.S. Fish and Wildlife Service. *Endangered Species. Defining Success Under the Endangered Species Act*. Accessed, 2013. https://www.fws.gov/endangered/news/episodes/bu-04-2013/coverstory/index.html.

United States General Accounting Office. *Endangered Species: Management Improvement Could Enhance Recovery Program*, Report to the Chairman, Subcommittee on Fisheries and Wildlife Conservation, and the Environment, Committee on Merchant Marine and Fisheries, House of Representatives. Washington, DC: GAO/RCED-89-5, December 1988.

U.S. General Accountability Office. Bureau of Land Management. *Effective Long-Term Options Needed to Manage Unadoptable Wild Horses*. Washington, DC: USGAO, GAO-09-77, 2008.

U.S. Senate. *Listing and Delisting Processes Under the Endangered Species Act*. Sect. Hearing before the Subcommittee on Fisheries, Wildlife, and Water of the Committee on the Environment and Public Works, 137th Congress First Session on the Regulations and Procedures of the U.S. Fish and Wildlife Service Concerning the Listing and Delisting of Species Under the Endangered Species Act, S. HRG. 107–322 May 9, 2001. www.gpo.gov/fdsys/pkg/CHRG-107shrg78073/html/CHRG-107shrg78073.htm.

Wildlife Management Institute. "Lessons from the Montana Wolf Management Stamp." *News Bulletin*, 68, 10 (October 2014). https://wildlifemanagement.institute/outdoor-news-bulletin/october-2014/lessons-montana-wolf-management-stamp.

Wilson, Edward O. "On *Silent Spring*." In *Courage for the Earth: Writers, scientists, and activists celebrate the life and writing of Rachel Carson*, edited by Peter Matthiessen, 27–36. Boston: Houghton Mifflin Company, 2007.

Wilson, Edward O. *Half-Earth: Our Planet's Fight for Life*. New York: Liveright Publishing Corporation, 2016.

Wilson, Patrick Impero. "Wolves, Politics and the Nez Perce: Wolf Recovery in Central Idaho and the Role of Native Tribes." *Natural Resources Journal* 39 (1999): 550–553. https://digitalrepository.unm.edu/cgi/viewcontent.cgi?article=1650&context=nrj.

Wisconsin Natural Resources Board. May 4, 2020. https://dnr.wi.gov/about/nrb/members.html.

Yachnin, Jennifer. "Does Biden's '30x30' Plan Trade Science for Popularity?" *E&E News*, June 2, 2021. https://www.eenews.net/greenwire/2019/08/06/stories/1060866919.

8 The Future of the Seascape and the Humanity of Islanders
Focusing on the Korean Archipelago[1]

Sun-Kee Hong

Islands all over the world have been changing physically and cognitively due to social, economic, and environmental transitions, both internal and external.[2] Such changes transform the characteristics of islands in time and space, depending on the magnitude and frequency of the change. In order to understand the relationship between the humanities-society-environment of the island and the sea, considering the interactions and effects on the daily lives of islanders in terms of the resources of islands and sea, and their surroundings,[3] we must identify the causes of various changes arising on the islands themselves and their surroundings, and study in depth how such changes affect the cognitive systems of islanders and outsiders.[4] It is also necessary to set up an academic scheme to understand the interconnections between the existence and life in islands; the space and vitality of islands; and the physical, psychological, and cognitive phenomena produced by changes to islands. Furthermore, we must reconceive and reorganize islands into visible and non-visible forms, shapes, or composition.[5]

This chapter aims to examine the interrelationship of knowledge, boundary, and network that indicate the connections in the main core axis representing the components of islands' humanities topography – including life culture, space, and community – and to explain what role these components play in the transition of humanities topography, and in the development of a sustainable island society, based on the characteristics of the Korean island region.

What Is the Island's Humanities Topography?

The term "changes in political topography" is used to refer to the political outlook and the fluctuation of political powers that appear according to changes in region, generation, and economic zones. Most of the Korean island area is a part of the continental shelf and so has geographical connectivity to the mainland. Of course, there are some islands, particularly those still in the process of formation, that are more isolated and have their own particular geographical characteristics. However, all of these islands have engaged in mutual cultural exchanges based on direct or indirect relationships with the mainland and connections with neighboring islands, regardless of any particular island's isolation.[6] The islands' economic and social relations with the continent have become closer as accessibility has increased in various ways with the advent of industrial society, and this trend is promoting the changes in the islands' unique cultural identity, that is, as Conkling argues, their islandness.

Since an island is surrounded by the sea and an isolated place far away from the continent or mainland, it may be thought of as a situation where a self-reliant economic structure is possible; however, many of the islands maintain a kind of subordinate relationship resulting from the continent's political and economic governance structure in modern times.[7] Therefore, even changes within the islands themselves are often determined by external circumstances. In this way, the world of the islands and islanders changes in response to the overall situation linked to the economy, society, politics, and culture of the continent (mainland), which can be interpreted as the term 'humanities topography'. In other words, if the concept of humanities topography is defined as "the aggregate shape of tangible and intangible experience, knowledge and lifestyle created by humans who settle in a specific place," the concept of island humanities topography can be defined as "the aggregate shape of tangible and intangible experience, knowledge and lifestyle created by humans who settle on an island, live on the sea as a space of livelihood, and interact with the land."

Humanities topography represents changes in reality; that is, the philosophical implication of each term is more like an amalgam than a simple combination of heterogeneous terms: humanities (life) and topography (space). Humanities topography has an evolutionary feature that changes, develops, or degenerates with the times. Humanities topography varies with changes in society, economy, and environment, and the phase is always flexible. Accordingly, it is necessary to emphasize that island humanities topography is in any phase of change and at any position always determined by the current reality of the islands. Consequently, the humanities topography of an island, which is changing due to social awareness and economic background, transforms the land on the island, and initiates a chain reaction that causes changes in patterns of natural resource use, which results in fundamental changes to the island's identity and constitutional tendency toward cultural diversity across the board.

What Does Humanities Topography Change?

Changes in islands' humanities topography are based on the perception of changes in response to the actual transitions on real islands (realism). Changes affecting the historical and ideological consciousness and views of the island world, accordingly, result in changes in an island's or islands' unique cultural identity. The result is a change in the visible–invisible aggregate figures that leads to a change in the island community. In some cases, an island's humanities topography is independently formed by the island community deriving its livelihood from the sea, which is a natural environment and a living space; conversely, a mainland perspective, or an external perspective may affect an island's humanities topography. Various sources of change – including change in island space from reclamation or other activities; natural disasters; and the larger economy and society – result in different changes of perception of the island, but the deep-rooted causes for such changes can be largely divided into the following comprehensive factors: society, economy, and environment.[8] It is necessary to reveal whether these causes occur inherently or externally.[9] Public perception of the islands is manifested in various experiences encompassing the islands and the sea, accumulated over a long period of time and passed down as traditional knowledge.[10] We can see the causes and effects of various changes from the evidence of human history.

Marine Awareness

Discussions of the public perception of islands has distinguished differences between countries and regions, and this distinction can be found in historical events. In the history of interchanges between continents and islands, as the perception of the ocean as an area has changed, so the perception of the neighboring sea also evolves, and eventually the new interest in the sea leads to the island. The political system and ruling power of the islands, the mainland, the inside of the continent and coastal area have changed the thoughts and awareness of neighboring islanders. Thus, the records left by the islanders represent a barometer of their awareness. Changes in an island's humanities topography may occur depending on the magnitude and breadth of spatiotemporal variability. The historical evidence since the islands' settlement – including development activities such as land clearance and reclamation, the number of settlers depending on the size of islands, and interactions with the neighboring islands – may allow us to identify the differential perceptions of islands, ocean, and marine space between islanders and mainland residents. In particular, marine awareness and identity may be understood through cultural exchange and trade via sea and islands, the world of sea peoples, which can be found in ancient and medieval history centered on Asia. How the world of island recognition in historical data and evidence currently affects modern times can also be seen as a form of change in humanities topography. Attempts have been made to consider the characteristics of the islands viewed from the perspective of the past continent from the position of island and islanders, which has been eventually set up as an academic field called "Nissology."[11]

Natural Environment

Factors that induce changes in the island's humanities topography are largely divided into external and internal factors, and external factors are classified into global and local scales according to their scale. One of the major factors on the global scale is the global climate crisis, which all maritime countries are facing internationally, including global warming,[12] an increase in seawater temperature, and rising sea levels.[13] This change in the marine environment of the earth results in melting glaciers in the Antarctic and Arctic, lowered ocean salinity, and the collapse of the marine ecosystem's food chain, which eventually degrades the global fishing environment. The local factors and their characteristics are determined by country and region. In general, they are divided into the sea areas of the West Sea, the Southwestern Sea, the South Sea, and the East Sea; the cultural features and identity of the island region are determined by their particular sea area.[14]

The island culture is largely influenced by ecological factors, including sea currents, ecosystem, resources, and other ecogeographical factors.[15] Geographical characteristics that function from the formation process of continents and ocean – including tidal mudflat, archipelago, marine islands, continental islands, rias coast, and sand dunes – eventually act as factors that determine modern living conditions. Changes in the environment of islands and ocean affect the overall island economy, including the livelihood and production of islanders. Therefore, it is necessary to discuss how the basic needs to live on the island have affected the cultural formation of the island society and what the modern meaning is. As the marine environment changes, the location of island villages, fishing grounds, port, and harbor also changes, and the islanders are demanding countermeasures to address such environmental changes. Changes in the natural environment will eventually become the fundamental cause of change in humanities topography. It is thus also essential to make a multidisciplinary and convergent interpretation of the interrelationship between different outcomes, ecological phenomena, and changes in the island space which appear from the changes in humanities topography.

Identity

Social prejudice against the islands has long persisted as a result of past history, but the perception of the islands has changed significantly in recent years. The reality of the islands was viewed from the perspective of others in the past, but the vividness of the island has been recently conveyed and disseminated in real time due to the practical and autonomous activities of islanders and the media. Now, the island research landscape is being changed by the islanders themselves. In particular, the national celebration, "the Day of Island," which was established in 2018, played an important role in raising awareness of the islands and improving interest in the island region among all people.

Island culture is created in the course of arriving on the island, settling the island, and interacting with other islands and the mainland; the hybridity and diversity appears in the course of such mixing. These cross-cultural negotiations generate a new culture, either assimilated with the old culture or else overtaking and eliminating the old culture. Indigenous beliefs, cultural landmarks (including holy sites), language, and food are characteristics of the island identity that emerge from both sides of the island's isolation and from new forms of communication. Islanders, who have been adapting to the sea, understand natural phenomena instinctively, detect changes quickly, and are finding alternative ways of adapting to them accordingly.[16] However, our thoughts and perceptions of the islands are still determined by others. Global trends in island identity research are also evolving significantly.

Science and Technology

Sustainability has been established through adaptation strategies, traditional knowledge, and the historical literature of the island. As island culture itself is a complex system that exhibits complex phenomena by means of various organic interactions between humans and nature; the method of reading variation should likewise be approached from a multilayered and interdisciplinary perspective. The island researchers, including humanities scholars, sociologists, and natural scientists, have been already discussing human "adaptation," "response," and "modification" to the rapidly changing environment in the Anthropocene.[17] Our islands and sea are different from those of the world, but the world's economic issues, climate, and environment already have a great influence on us. Will the currently existing knowledge and information about the islands suffice for future islands in determining their value? Indeed, what is the future value of the islands for the islanders? An era of operating all aquafarms and farms in a remote-controlled manner is coming.

Knowledge, Boundary, Network: Components of Humanities Topography and Their Function

The question of "From what evidence do we read the changes in the island's humanities topography?" will eventually be narrowed down to "what do we look at on the island?" Since the breadth of ideological views of the islands is infinite, the concept of humanities topography, the breadth and depth of variation, and the indicators of change are determined depending on "where our view will be placed."

Knowledge, boundary, and network are aspects of changes in humanities topography, as well as the components of humanities topography. Knowledge, boundary, and network can be explained as the linkage of "communication and interaction" between "human (life)," "space (place)," and the sea. The island's humanities topography has the peculiarity of an interrelationship in which these three elements are interconnected or separated by boundaries (between islanders, mainlanders, and the sea area). The past communication and interaction via the sea, including wisdom to read nature, science and technology to cross the sea, exchange with other cultures for trade, and coexistence and competition with other peoples, represent knowledge that drives the future of the islands and the sea.[18] This past knowledge is information that is essential for driving a sustainable future.

Modern islands are challenged by a number of new factors. The island society is changing due to intrinsic factors such as aging and population reduction.[19] A variety of projects are also conducted to attract tourists.[20] There are also practical projects to be developed for future generations, such as improving welfare of islanders, making efforts for the island's self-reliant economy, responding to the climate crisis, and preserving clean water,[21] which is required by the islanders.

Awareness of future situations is also necessary for rebuilding and restoring island culture. Furthermore, the three components of knowledge, boundary, and network must be addressed in the study of island humanities, with an effort to anticipate the direction in which they will move. The sustainability of islands will be maintained only when knowledge, boundary, and network in the island society are driven by a humanistic-interacting web that is not simply linear and enumerative, but mutually organic and ecologically multi-functional.[22]

Knowledge: Island, a Museum of Marine Knowledge

Knowledge is a unique cultural gene of an era, which is passed down from generation to generation. Knowledge includes a traditional cognitive system that humans have adapted to use limited resources in a barren natural environment, and is passed down to future generations as a tool for future survival.

The typical livelihood in an island and marine area, harvesting and fishing, encompasses traditional knowledge with ecological and cultural value.[23] Simple tools, harvesting and fishing techniques, social systems and indigenous knowledge thus far passed down are playing a role in obtaining highly efficient survival resources and in maintaining ecological balance through minimal human intervention in natural process. In view of these facts, the knowledge system in the island region can only be understood by including both the humans and the culture of the entire ecosystem (e.g. academia uses the term "landscape" in a broader sense rather than "ecosystem"), and then by taking an ecosystemic approach to the entirety.[24] This approach also offers a way to preserve island knowledge and creatively pass down these cultural factors, to prepare for modern and future living environment problems, and to meet ecological conservation, economic needs, and cultural diversity in a harmonious manner.[25]

Traditional knowledge develops into future knowledge. Korea's current science and technology related to fishery and aquaculture has reached a level that can restore the unique ecosystem of the community and be used in an environmentally friendly and ecological manner. In cultural terms, human, social, and cultural conditions must be created to activate the autonomous organizations of residents, to revive traditional and cultural elements, and to set up ecological fishing and aquaculture. Significantly, this sort of positive cultural and environmental evolution will not occur on its own, since modern culture and economic motives otherwise tend toward exploitation of resources.

Boundary: Island on the Boundary

Boundaries can appear as psychological, physical, or cognitive.[26] The nature of the sea that separates the continent from the islands instills psychological boundaries for both islanders and mainlanders. The psychological stability of the boundary also varies depending on the distance between the continent and the island and the size of the island. The existence of an island in the boundless and vast ocean offers a kind of psychological relief.

For both islanders and land residents, the sea is a boundary that puts them in the position of "others" to each other. The loneliness and solitude that tourists from the mainland feel when they arrive on an island, as well as the curiosity and sense of superiority in the corner of their hearts, may represent the boundary arising from the perception of "mainlanders." Sometimes, this perception of boundary stems from economic and political "power." It represents a mindset of trying to control – the larger islands tend to own, control, and use the smaller islands, and the mainland tends to own, control, and use the island region.

The construction of bridges is a long-cherished project for residents who wanted to link the islands and the mainland to be able to leave the islands, but, on the other hand, bridges also represent an investment in "land capital." The psychological and physical boundary between an island and the mainland may be temporarily removed by a huge bridge, but the island is in fact an eternal island.

The debris left by mankind, including marine waste, radioactive contamination, and micro-plastics, moves through the sea to form new boundaries. Like fish, this debris settles down while traversing the ocean freely without recognizing borders or regions. When the European Empire was scrambling to acquire resources in the Age of Exploration, in its colonial conquest of Asia, the Pacific, and South America, goods and people as well as animals, plants, and diseases also moved together. Therefore, an island is both a stepping-stone connecting the continent and the sea, and also a settlement.

Network: Openness, Self-reliance, and Democracy

The study of and concern for an island's identity (islandness) and diversity (characteristics) are common among island researchers all over the world and represent the fundamentals of island research. When discussing an island's identity, it can be said that the ever-present concern is the dual nature of any island: isolation and communicability. "Isolation" reflects geographical recognition characteristics of understanding an "island" as an independent formative body lacking connectivity due to the environmental backdrop of the sea. "Communicability" is a positivity exhibited in the active behavior of human beings which arises from travel back and forth between island and mainland as well as immigration to an isolated island environment for various purposes and reasons. Just as humans drift through the sea and islands, or land on and live on islands, travel between and migration from island to island also apply to various marine lives. The most important material in discussing the island's identity is the physical and psychological features of communication (interaction) and isolation, but the economic disturbance, cultural hybridization, and natural phenomena surrounding resources also play important roles. Thus, communication and isolation are often like two faces of the same coin, rather than simply representing their respective dichotomous concepts.

The sea area (water area) plays a very important role as a medium for spreading living things and culture in understanding the island-to-island, island-to-continent connectivity, and the nature of this network.[27] The role of the sea area is to act as buffer that ensures stability of boundaries and improves accessibility to control and supplement the ratio of communication to isolation.

Depending on the relative accessibility to land,[28] the presence or absence of a bridge, and active marine economy, the nature of a network appears more or less dense. On the other hand, the nature of such complex and dense "networks" plays a role in diminishing the unique characteristics of the island. In particular, the structures that forms the origin of cultural resources, such as language and tribe, are sometimes used as indicators for evaluating networkability, which can be addressed in terms of biodiversity transmitted in the form of traditional knowledge.

Interdisciplinary Approach for Research Methodology of Island Humanities Topography

New research based on the existence, identity, and future-oriented direction of islands is required, including research focused on the concept of islandness and the relative share of openness and interaction of an island determined by differences in geographical, topographical, and spatial distribution, along with the maritime thinking and the existing insularity based on the continental perception of the islands.[29] All of the various environmental, socio-economic, and identity problems of the islands in Korea have similarities to those of other islands around the globe. In particular, the academic lacuna in not yet considering the sustainability of islands and the activation of island communities imperils the future like a fragile house of cards, by overlooking changes in the island and ocean culture due to excessive human activities, including changes in ecosystems caused by climate change, collapse of communities caused by the decline of traditional fishing, and the damage to cultural relics caused by tourism development. All these problems are realistic and common to the islanders all over the world. How do we solve these problems in a humanities way, or else how do we integrate them with other studies and make creative suggestions?

If possible, research on target islands should be conducted, focusing on the types of islands that are differently influenced at the physical and socio-economic level, such as isolated islands, archipelagos, and the land-linked islands; such research should be academically generalizable and also provide a positive feedback to the island society to help spread the results. In order to carry out this new research, the method of "consilience," based on the wide receptivity of the humanities itself will have to open the chapter of convergence with other academic area.[30] It is important to consider "what is on the island," and "from where we get what," but the island researcher will also need to develop an agenda to think about "what we will feed back to the island."

Common Interests and Collaboration with Other Disciplines

Islands vary in geographical and morphological classifications, and for inhabited islands, the lifestyle of islanders varies depending on the characteristics of the surrounding marine ecosystem and biological resources. It is very difficult to select an island that can be studied by a group of experts from various fields, especially considering its historical background and relationship with the mainland after landing on and settling the island.[31] However, by comparing the results of research on island identity and ecological culture, such as livelihood and the use of resources in Korea's archipelago region, we may classify the research into various areas of study, delineated by five keywords. The first is to study the characteristics of island itself and relationship with the surrounding sea water such as space, resource and sea area; the second is to reveal social relationships, such as island village or fishing communities; the third is the method and techniques for the use of land, fisheries, and biological resources; the fourth is the traditional ecological knowledge for understanding natural phenomena and trends related to climate

Table 8.1 Research topic and academic fields, focusing on keywords available for collaboration

Keywords	Classification
Diversity	Biodiversity, cultural diversity, biocultural diversity, diversity of resources and production, ecosystem, ecological culture
Space	Spatial recognition, island regeneration, landscape, sea area, interaction, climate change, space change due to landfill and land use, etc.
Knowledge	Traditional ecological knowledge, future knowledge, knowledge tradition, transition of traditional knowledge, changes in marine ecosystems and fishery resources, fishing villages
Islandness	Island–continent relation theory, democracy, independence, island links to the mainland, changes in islanders' consciousness caused by increased accessibility, culture shock by tourism
Sustainability	Balanced development, community society, livelihood, industry, tourism, ecosystem service, disaster (natural, artificial), future society

(wind, temperature, and wave height), such as ocean currents, tides and constellations; and the final is humanities stories, such as history, faith, and rituals of the island.[32] In recent years, since various policies on islands have been actively carried out in Korea, and national awareness for islands is increasing, through activities such as the designation of the national celebration day, "the Day of Island," the direction of the island policy would be also included as the sixth keyword in terms of preserving and inheriting the island's unique ecological culture.

Table 8.1 summarizes the six collaborative research areas presented above in five keywords. Each of these topics is commonly considered by island researchers all over the world, as well as an essential research topic that is considered important in determining the direction for researching an island's humanities topography in Korea. For these five points, the researchers can creatively cooperate with other academic fields.

Diversity

It is necessary to discuss the biological and cultural diversity that may arise depending on the distribution, shape, size, and geographical characteristics of an island.[33] The culture of island and fishing villages is inevitably developed by islanders who rely on biological resources, and, therefore, the biological and cultural diversity is related to mutual dependence, symbiosis, and coevolution. Biocultural diversity is a complex characteristic caused by living things, ecosystem, and race[34]; therefore, rituals, indigenous languages, the use of biological resources (fermentation, medicinal herbs, etc.), food (food ingredients), residential areas, and fishing methods can be seen as very important biocultural resources. Since biology and culture are different attributes, but humans depend on nature (having repeatedly developed and been in a fateful reliance such that they cannot exist without the use of natural resources), it is believed that the term "biocultural diversity" was created to mean the coexistence between nature and humans.[35]

Biocultural diversity is manifested in the dynamic process in which interactions between biodiversity, cultural diversity, and traditional knowledge operate in a complex ecological system.[36] Humans have used the surrounding landscapes and living things as living resources for a long time and, as necessary, have developed new species through cultivation. The use of biodiversity has become the backdrop for promoting cultural diversity such as food culture and residential culture, and this ecological knowledge has spread beyond adjacent regions to the national level. In addition, local languages and dialectics are very vulnerable to Westernization and are rapidly disappearing. Indigenous knowledge of the use of natural resources is facing a crisis of biodiversity driven by reckless energy development and land use.[37]

It is undeniable that humans and nature have mutually relied on each other, as connected through and complemented by an ecosystem, but governments, researchers, citizens, and experts need urgently to understand that this connection has been declining due to rapidly changing global environment, reckless development, and reduced biodiversity. As seen in human history, the survival of humans in the future will depend significantly on biodiversity; moreover, the ecocultural flexibility and sustainability shown in these interrelationships between biodiversity and cultural diversity will be used as a model for harmonious coexistence with ecosystems that can support the existence of human beings in the future.

Space

It is necessary to study the past history and current status of islanders' recognition of ancestors who initially settled an island – people or family who settled first and started living on a specific island, who formed a kind of clan society in early days. It is also necessary to study how the spatial perception of islanders based on historical evidence from the past differs from such special perception at the present moment, and how the present sense of space differs from that of the past. The use of space by islanders is changing as various commercial and passenger ships are operated, bridges are constructed, and accessibility is improved.[38] It is a reality that the recent climate changes have caused many complaints from islanders who now demand changes in the location of existing ports. Space is a datum that can provide information about perception in the past and at present as long as there is a record. Nevertheless, a more quantitative and qualitative analysis of drawing data is required.[39] Accessibility to the islands is improving compared to the past. Thus, the exchange of materials and information has become faster and more accurate. Meanwhile, some portion of the island's population has moved to the mainland because of convenient transportation, and the number of islanders living on both the islands and the mainland is increasing. These dual residents spend time in mainland cities in winter, and on islands in summer. As such, the perception of the space understood as an island is increasingly changed by such physical changes involving the islands themselves.

Knowledge

It has already been revealed by many researchers that much crucial information, which can be said to be the wisdom of life, has been passed down to each islander. The knowledge on the island is so important that an elderly resident of the island is called an "island museum." Traditional knowledge, both tangible and intangible – such as wind, tide time, fishing gear, fishery, farming methods, and biological use (medicinal, edible) can be said to be a common heritage of mankind.[40]

Digitalization of traditional knowledge, and the incorporation of traditional knowledge into future knowledge are essential in the development of sustainable islands. Knowledge, which had been passed down only to the islanders and within communities, is becoming widespread as humankind becomes an information culture.[41] In other words, the indigenous knowledge of islanders and mariners who have been adapted to the changes of barren natural environment will help create a new future industry that can lead to the realization of sustainability by combining cutting-edge response strategies, including science and technology required for the survival of future human beings who are in danger of climate crisis.

Islandness

An island's unique topographical and geological characteristics, and the characteristics different from the mainland (or similar characteristics), are manifested as identity.[42] The study of the characteristics of marine islands and continental islands, archipelagos (as well as the relationship between the main island and sub-islands in the archipelago) will reveal an island's unique identity, including its island space, perception of the sea, resource specificity, and self-reliance. Some say "the island is closed," and others argue "the island is open," because it faces the sea. Closure and openness are two-sided characteristics of an island, and such double-sidedness represents its islandness. An island is both closed and open.[43] In particular, such islandness is differentiated by what surrounds the island. Are those surroundings a tidal mud flat, another island, or a boundless vast ocean? The background of defining an island – such as whether it is large or small, adjacent to land, part of an archipelago – varies greatly depending on the island, the surrounding sea, and its geopolitical position.[44] However, what directly or indirectly affects the islandness is the mentality of the islanders. In other words, it is believed that the sociality of those who stepped on the island first is extremely relevant. As there is a saying of neighborhood hearts and mind, warm-heartedness and emotions vary from island to island. Just as genes vary from person to person, the characteristics of islandness differ according to the combination of people who formed the island society.

Sustainability

Recently, the concept of "fairness" is spreading nationwide. However, following Melvin Lerner's advice, "Don't be buried in a fair world hypothesis," it is necessary to examine carefully whether or not the island reality can be a suitable model for a "fair" society. Sustainability is a kind of process in which social, economic, and environmental systems develop together in a balanced manner. Although there are no clearly defined sustainability goals, the UN's recent sustainable development goals (SDGs) described many protocols in relation to the developmental direction of islands around the world, the maintenance of living standards in islanders' communities, and the response to climate change. It is also essential to study how to apply this international and domestic direction to island areas and maintain the quality of life and economy of the islanders.

Islanders' sustainability is possible when an island's environmental ecological system, the "biosphere," and human social system "culture" coexist and balance, but there is still a significant lack of discussion on assessment and indicators for sustainability of island regions in the Asia-Pacific, such as in Korea. Considering that there is an urgent need to improve the living environment in island regions vulnerable to global climate change[45] – including the rise of sea levels due to changes in marine climate, changes in the infrastructure of farming and fishing resulting from changes in the island environment, and natural disasters such as earthquake and tsunamis – it is necessary to preserve ecosystems in the coastal area, build infrastructure for the islanders, and implement a qualitative economic system. It is also essential to preserve and utilize intrinsic biological and cultural resources of the archipelago on the West and South Sea, Korea's representative island area, to improve the quality of life of islanders, and to share their ecological values with people all over the world as we learn to be better stewards of islands the world over.

Conclusion

Islandness is an ambiguous concept that is partially derived from the simultaneous openness and closure of the island boundary. Openness suggests a connection to a wider world, and closure relates to insularity – disregard or neglect of people, culture, and ideas outside of one's own experience. This ambiguity becomes apparent in the "tension" that arises between the islanders' desire for self-governance and their desire for equality with the mainlanders. This sense of islandness may be weakened by increased accessibility – in the form of greater boundary opening, such as bridges or ferries – or by the cost-free development resulting from sustainability or the specific characteristics of an island. But insularity may also be a problem. Many offshore islands must be open to tourism to maintain economic viability. The key question is how to balance the obvious need for future economic development – to have homogeneity with the mainland – and the need to maintain the island's unique characteristics.

Boundaries may appear as psychological, physical, or cognitive. The existence of the boundary between island and mainland makes an island represent both an imaginary ideal and a geographic possibility. An island may invoke a desperate sensation among islanders that they can never escape, but for mainlanders, it represents an escape to a new open space. At any time, island development and interest begin with mainland-based thinking. Throughout the era of modernization, the Korean archipelago has become a land of development. Smaller islands were reclaimed to create larger islands, tidal flats were reclaimed to form land, and salt fields and cultivated lands were built on them. The inhabitants of the islands want as much access to the mainland as possible, and they have long hoped for a connection to the mainland. Today, such a wish leads to the construction of island–land bridges. The desires of these islanders are then invoked by mainland capitalists for their own development agendas. Islands as points connect to a line and become an area. Not only the physical connection between the point–line–area but also the psychological and imaginary connections are implicit in the connection between islands and the mainland.

The knowledge transmitted by the islanders is transformed by the continuous physical and human networking on the island. The technological tradition and many biological cultures that harvested fish in the traditional way are now gone. Unfortunately, islands – especially the Korean islands – are experiencing social problems due to rapid aging and declining populations. Even though the problems that the islands are experiencing must be solved for the islanders, the island is usually only considered and viewed as a relative to the mainland; its own existence is seen as merely incidental. Our current understanding of islands is one in which traditional knowledge and new social challenges must be brought together as we search for solutions in stewarding both the future of islands and the future of the planet.

Notes

1. Thank you to Professors Rachel Carnell and Chris Mounsey for inviting me to participate in this project. Further, I would like to express my gratitude to Professor Carnell for various editorial suggestion on my manuscript and to Professor Mounsey with his assistance in formatting. This manuscript was carried out with the support of the Humanities Korea Plus (HK+) project of the National Research Foundation of Korea (2020S1A6A3A01109908).
2. Godfrey Baldacchino, "How Far Can One Go? How Distance Matters in Island Development," *Island Studies Journal* 15, no. 1 (2020): 25–42; Andreas Østhagen, "A Sea of Conflict? The Growing Obsession with Maritime Space," accessed September 7, 2021, *The Arctic Institute*, February 12, 2019, https://www.thearcticinstitute.org/sea-conflict-growing-obsession-maritime-space/.
3. Philip Conkling, "On Islanders and Islandness," *Geographical Review* 97, no. 2 (2007): 191–201.
4. Christian Depraetere, "The Challenge of Nissology: A Global Outlook on the World Archipelago Part 1: Scene Setting the World Archipelago," *Island Studies Journal* 3, no. 1 (2008): 3–16; Christian Depraetere and Arthur Dahl, "Island Locations and Classifications," in *A World of Islands, An Island Studies Reader*, ed. Godfrey Baldacchino (Malta & Canada, Agenda Academic and Institute of Island Studies, 2007): 57–105.
5. Sun-Kee Hong, Jan Bogaert and Qingwen Min, *Biocultural Landscapes: Diversity, Functions and Values* (Dordrecht: Springer, 2014); Sun-Kee Hong, et al., "Interdisciplinary Convergence Research Design on Island Biocultural Diversity-Case Study in Wando-gun (County) Island Region, South Korea," *Journal of Marine and Island Cultures* 7, no. 1 (2018): 12–37.
6. Grant McCall, "Nissology: A Proposal for Consideration," *Journal of the Pacific Society* 17, nos. 2–3 (1994): 1–14.
7. Hiroshi Kakazu, *Sustainable Development of Small Island Economics* (Boulder: Westview Press, 1994).
8. Hong, et al., "Interdisciplinary Convergence Research Design on Island Biocultural Diversity."
9. Kakazu, *Sustainable Development*.
10. Hong et al., "Interdisciplinary Convergence Research Design on Island Biocultural Diversity."
11. McCall, "Nissology"; Depraetere, "The Challenge of Nissology."
12. Paul J. Crutzen, and Eugene F. Stoermer, "The Anthropocene," *Global Change Newsletter* 41 (2000): 17.
13. AR6, "Climate Change 2022: Impacts, Adaptation and Vulnerability," accessed September 7, 2021, https://www.ipcc.ch/report/sixth-assessment-report-working-group-ii/.
14. Sun-Kee Hong et al., "Challenges and Goals of the Sustainable Island: Case Study in UNESCO Shinan Dadohae Biosphere Reserve, Korea," in *Designing Low Carbon Societies in Landscapes*, eds. Nobukazu Nakagoshi and Jhonamie A. Mabuhay (Tokyo: Springer, 2014).
15. Samantha Chisholm Hatfield and Sun-Kee Hong. "Mermaids of South Korea: Haenyeo (Women Divers) Traditional Ecological Knowledge, and Climate Change Impacts." *Journal of Marine and Island Cultures* 8, no. 1 (2019): 1–16.
16. Clifford Geertz, *Local Knowledge: Further Essays in Interpretive Anthropology* (New York: Basic Books, 1983), 7; Fikret Berkes, "Indigenous Ways of Knowing and the Study of Environmental Change," *Journal of the Royal Society of New Zealand* 39, no. 4 (2009).

17 Clarence Alexander et al., "Linking Indigenous and Scientific Knowledge of Climate Change." *BioScience* 61, no. 6 (2011): 477–484.
18 Johanna Johnson, et al., "Priority Ddaptations to Climate Change for Pacific Fisheries and Aquaculture: Reducing Risks and Capitalizing on Opportunities. FAO/Secretariat of the Pacific Community Workshop, 5–8 June 2012, Noumea, New Caledonia," *FAO Fisheries and Acquaculture Proceedings* 28 (2013).
19 Sun-Kee Hong, "Local Activation Using Traditional Knowledge and Ecological Resources of Korean islands," *Journal of Ecology and Environment* 38, no. 2 (2015): 263–269.
20 Sun-Kee Hong and Nobukazu Nakagoshi, *Landscape Ecology for Sustainable Society* (Cham: Springer, 2017); Gang Hong, "Islands of Enclavisation: Eco-Cultural Island Tourism and the Relational Geographies of Near-Shore Islands," *Area* 52, no. 1 (2020): 47–55.
21 Kakazu, *Sustainable Development*.
22 Hong et al., "Challenges and Goals of the Sustainable Island" in Hong and Nakagoshi. *Landscape Ecology for Sustainable Society*.
23 Berkes, "Indigenous Ways of Knowing and the Study of Environmental Change."
24 Hong, et al., *Biocultural Landscapes*.
25 Clarence Alexander, et al., "Linking Indigenous and Scientific Knowledge of Climate Change," *BioScience* 61, no. 6 (2011): 477–484.
26 Depraetere, "The Challenge of Nissology"; Soung-Su Kim, "An Examination of Five Terms Related to the Phenomenon of the Trans-boundaries of Culture: Consilience, Fusion, Convergence, Complex, and Articulation." *Philosophy and Culture* 26 (2013); Tchi-Wan Park, "Philosophy on the Border between Locality and Globality, and Relativity and Universality." Philosophy and Culture 26 (2013).
27 Kakazu, *Sustainable Development*.
28 Adam Grydehøj and Marco Casagrande, "Islands of Connectivity: Archipelago Relationality and Transport Infrastructure in Venice Lagoon." *Area* 52, no. 1 (2020).
29 Sun-Kee Hong and Gloria Pungetti, "Marine and Island Cultures: A Unique Journey of Discovery." *Journal of Marine and Island Cultures* 1, no. 1 (2012): 1–2.
30 See also Claude Lévi-Strauss's idea of "Bricolage" from *The Savage Mind* (London: Weidenfeld and Nicholson, 1962).
31 Hatfield and Hong, "Mermaids of South Korea: Haenyeo (Women Divers) Traditional Ecological Knowledge, and Climate Change Impacts."
32 Kyong-Yeop Lee, *Folklore of the Island, Read with Four Keywords* (Seoul: Minsokwon Archebooks, 2020).
33 Jose Maria Fernández-Palacios, Christoph Kueffer and Donald R. Drake, "A New Golden Era in Island Biogeography," *Frontiers of Biogeography* 7, no.1 (2015).
34 Luisa Maffi and Ellen Woodley, *Biocultural Diversity Conservation – A Global Sourcebook* (London: Earthscan, 2010).
35 Sun-Kee Hong, "Eco-Cultural Diversity in Island and Coastal Landscapes: Conservation and Development," in *Landscape Ecology in Asian Cultures*, eds. Sun-Kee Hong, Jianguo Wu, Jae-Eun Kim, Nobukazu Nakagoshi (Tokyo: Springer, 2011).
36 Pretty Jules, et al., Plenary paper for Conference "Sustaining Cultural and Biological Diversity in a Rapidly Changing World: Lessons for Global Policy," organized by *The American Museum of Natural History's Center for*

Biodiversity and Conservation, IUCN-The World Conservation Union/Theme on Culture and Conservation, and Terralingua (April 2–5 2008).
37 David Rapport, "Sustainability Science: An Ecohealth Perspective." *Sustainability Science* 2, no. 1 (2006).
38 Grydehøj and Casagrande. "Islands of Connectivity."
39 Hong et al., "Interdisciplinary Convergence Research Design on Island Biocultural Diversity."
40 Geertz, *Local Knowledge: Further Essays in Interpretive Anthropology*; Berkes, "Indigenous Ways of Knowing and the Study of Environmental Change."
41 Chisholm Hatfield and Hong. "Mermaids of South Korea."
42 Conkling, "On Islanders and Islandness."
43 Hong Pungetti, "Marine and Island Cultures: A Unique Journey of Discovery."
44 Hong, "Eco-Cultural Diversity in Island and Coastal Landscapes: Conservation and Development."
45 IPCC, 2001, "Small Island States," *Climate Change 2001: Working Group II: Impact, Adaptation and Vulnerability* (Cambridge: Cambridge University Press, 2001).

Bibliography

Alexander, Clarence, Nora Bynum, Elizabeth Johnson, Ursula King, Tero Mustonen, Peter Neofotis, Noel Oettlé, Cynthia Rosenzweig, Chie Sakakibara, Vyacheslav Shadrin, Marta Vicarelli, Jon Waterhouse and Brian Weeks,. "Linking Indigenous and Scientific Knowledge of Climate Change." *BioScience* 61, no. 6 (2011): 477–484.
AR6, "Climate Change 2022: Impacts, Adaptation and Vulnerability." Accessed 7 September 2021. https://www.ipcc.ch/report/sixth-assessment-report-working-group-ii/.
Baldacchino, Godfrey. "How far can one go? How distance matters in island development." *Island Studies Journal* 15, no. 1 (2020): 25–42.
Berkes, Fikret. "Indigenous ways of knowing and the study of environmental change." *Journal of the Royal Society of New Zealand* 39, no. 4 (2009): 151–156.
Chisholm Hatfield, Samantha and Sun-Kee Hong. "Mermaids of South Korea: Haenyeo (Women Divers) Traditional Ecological Knowledge, and Climate Change Impacts." *Journal of Marine and Island Cultures* 8, no. 1 (2019): 1–16.
Conkling, Philip. "On Islanders and Islandness." *Geographical Review* 97, no. 2 (2007): 191–201.
Crutzen, Paul J., and Eugene F. Stoermer. "The Anthropocene." *Global Change Newsletter* 41 (2000): 17–18.
Depraetere, Christian. "The Challenge of Nissology: A Global Outlook on the World Archipelago Part 1: Scene Setting the World Archipelago." *Island Studies Journal* 3, no. 1 (2008): 3–16.
Depraetere, Christian, and Arthur Dahl. "Island Locations and Classifications." In *A World of Islands, An Island Studies Reader*, edited by Godfrey Baldacchino, 57–105. Malta & Canada: Agenda Academic and Institute of Island Studies, 2007.
Depraetere, Christian & Arthur, Dahl, 2018. 'Locations and Classifications' in Godfrey Baldacchino (ed.) The Routledge International Handbook of Island Studies, A World of Islands, pp. 21–51.

Fernández-Palacios, Jose Maria, Christoph Kueffer and Donald R. Drake. "A New Golden Era in Island Biogeography." *Frontiers of Biogeography* 7, no. 1 (2015): 1–7.

Geertz, Clifford. *Local Knowledge: Further Essays in Interpretive Anthropology*. New York: Basic Books, 1983.

Grydehøj, Adam, and Marco Casagrande. "Islands of Connectivity: Archipelago Relationality and Transport Infrastructure in Venice Lagoon." *Area* 52, no. 1 (2020): 56–64.

Hong, Gang. "Islands of Enclavisation: Eco-cultural Island Tourism and the Relational Geographies of near-shore islands." *Area* 52, no. 1 (2020): 47–55.

Hong, Sun-Kee. "Biocultural Diversity and Traditional Ecological Knowledge in Island Regions of Southwestern Korea." *Journal of Ecology and Field Biology* 34, no. 2 (2010): 137–147.

Hong, Sun-Kee. "Eco-Cultural Diversity in Island and Coastal Landscapes: Conservation and Development." In *Landscape Ecology in Asian Cultures*, edited by Sun-Kee Hong, Jianguo Wu, Jae-Eun Kim, Nobukazu Nakagoshi, 11–28. Tokyo: Springer, 2011.

Hong, Sun-Kee. "Local Activation Using Traditional Knowledge and Ecological Resources of Korean Islands." *Journal of Ecology and Environment* 38, no. 2 (2015): 263–269.

Hong, Sun-Kee and Gloria Pungetti. "Marine and Island Cultures: A Unique Journey of Discovery." *Journal of Marine and Island Cultures* 1, no. 1 (2019): 1–2.

Hong, Sun-Kee and Nobukazu Nakagoshi, eds. *Landscape Ecology for Sustainable Society*. Cham: Springer, 2017.

Hong, Sun-Kee, Heon-Jong Lee, Bong-Ryong Kang, Jae-Eun Kim, Kyoung-Ah Lee, Kyoung-Wan Kim and Dae-Hoon Jang. "Challenges and Goal of the Sustainable Island: Case Study in UNESCO Shinan Dadohae Biosphere Reserve, Korea." In *Designing Low Carbon Societies in Landscapes*, edited by Nobukazu Nakagoshi and Jhonamie A. Mabuhay, 145–162. Tokyo: Springer, 2014a.

Hong, Sun-Kee, Jan Bogaert and Qingwen Min. *Biocultural Landscapes: Diversity, Functions and Values*. Dordrecht: Springer, 2014b

Hong, Sun-Kee, Yong-Tae Won, Gyeong-A Lee, Eun-Seon Han, Mi-Ra Cho, Hye-Yeong Park, Jae-Eun Kim and Samantha Chisholm Hatfield. "Interdisciplinary Convergence Research Design on Island Biocultural Diversity-Case Study in Wando-Gun (County) Island Region, South Korea." *Journal of Marine and Island Cultures* 7, no. 1 (2018): 12–37.

IPCC, 2001 "Small Island States." In *Climate Change 2001: Working Group II: Impact, Adaptation and Vulnerability*. Cambridge: Cambridge University Press, 2001.

Johnson, Johanna, Johann Bell and Cassandra De Young. "Priority Adaptations to Climate Change for Pacific Fisheries and Aquaculture: Reducing Risks and Capitalizing on Opportunities. FAO/Secretariat of the Pacific Community Workshop, 5–8 June 2012, Noumea, New Caledonia." *FAO Fisheries and Acquaculture Proceedings* 28 (2013).

Kakazu, Hiroshi. *Sustainable Development of Small Island Economics*. Boulder: Westview Press, 1994.

Kim, Soung-Su. "An Examination of Five Terms Related to the Phenomenon of the Trans-boundaries of Culture: Consilience, Fusion, Convergence, Complex, and Articulation." *Philosophy and Culture* 26 (2013): 115–144 (Korean).

Lee, Kyong-Yeop. *Folklore of the Island, Read with Four Keywords*. Seoul: Minsokwon Archebooks, 2020 (Korean).

Lévi-Strauss, Claude. *The Savage Mind*. London: Weidenfeld and Nicolson, 1962.

Maffi, Luisa and Ellen Woodley. *Biocultural Diversity Conservation – A Global Sourcebook*. London: Earthscan, 2010.

McCall, Grant. "Nissology: A Proposal for Consideration." *Journal of the Pacific Society* 17, nos. 2–3 (1994): 1–14.

Østhagen, Andreas. "A Sea of Conflict? The Growing Obsession with Maritime Space." The Arctic Institute. February 12, 2019. Accessed September 7, 2021. https://www.thearcticinstitute.org/sea-conflict-growing-obsession-maritime-space/.

Park, Tchi-Wan. "Philosophy on the Border between Locality and Globality, and Relativity and Universality." *Philosophy and Culture* 26 (2013): 41–75 (Korean).

Pretty, Jules, Bill Adams, Fikret Berkes, Simone Ferreira de Athayde, Nigel Dudley, Eugene Hunn, Luisa Maffi, Kay Milton, David Rapport, Paul Robbins, Colin Samson, Eleanor Sterling, Sue Stolton, Kazuhiko Takeuchi, Anna Tsing, Erin Vintinner and Sarah Pilgrim. Plenary paper for Conference "Sustaining Cultural and Biological Diversity in a Rapidly Changing World: Lessons for Global Policy." Organized by *The American Museum of Natural History's Center for Biodiversity and Conservation*, IUCN-The World Conservation Union/Theme on Culture and Conservation, and Terralingua. April 2–5, 2008.

Rapport, David. "Sustainability Science: An Ecohealth Perspective." *Sustainability Science* 2, no. 1 (2006): 77–84.

9 Stewardship of Rangelands in the 21st Century

Managing Complexity from the Margins

Nathan F. Sayre

The term *range* has an Old World etymology, but its meaning cannot be separated from European expansion, conquest, and settlement. According to the *Oxford English Dictionary*, *range* as a noun derives from the Old French verb *renger*, which denoted the movement of herders and livestock across extensive open areas of land. It dates from the late fifteenth century, just prior to Columbus's voyages to the Americas. Many of the lands into which Europeans would subsequently expand, notably North America and Australia, lacked domesticated livestock, so, strictly speaking, they also had no rangelands – until Europeans arrived with their cattle, goats, sheep, horses, and pigs, that is. And as Alfred Crosby has shown, domesticated livestock were among the Europeans' most powerful weapons of ecological imperialism, not simply for the work they could do or the food they could provide, but above all because indigenous peoples lacked immunity to the many Old World diseases – smallpox, measles, mumps, and influenza, for example – that had evolved from the prolonged proximity of people and livestock at high densities. "It was their germs," Crosby writes, "not these imperialists themselves, for all their brutality and callousness, that were chiefly responsible for sweeping aside the indigenes and opening the Neo-Europes to demographic takeover."[1] Richard White puts the matter bluntly: "Without domesticated animals, Europeans would have neither survived nor conquered" in the New World.[2] In more ways than one, then, range livestock production was "the principal means whereby Europeans colonized and exploited the natural resources of sub-Saharan Africa, Australia, North and South America."[3]

The definition of range – and rangeland, which is the preferred term nowadays – has changed over time, but its frontier underbelly persists. In current usage, rangelands are a type of land, or rather a collection of types of land: grasslands, prairies, savannas, shrublands, steppes, tundra, and deserts. Each of these can plausibly be understood as a kind of biome or ecosystem, but the same cannot be said for the encompassing category, rangelands, since the constituent elements do not share any biologically or ecologically relevant attribute. All they have in common is that they are *not* any of the other major types employed to classify Earth's land

DOI: 10.4324/9781003219064-13

cover. Rangeland is thus a residual category, a catch-all for any landscape that is neither forested, cultivated, buried in ice, built up, nor paved over. From an evolutionary perspective, livestock enabled Old World peoples to secure reliable livelihoods from non-arable landscapes by converting natural vegetation into edible calories and protein; from a modern perspective, rangelands are places not (yet) put to some other, more intensive use, and therefore (still) available for livestock grazing, expressly or as a kind of default or placeholder land use.

Little wonder, then, that rangelands are shrinking as humanity gradually converts them to other purposes. The Millennium Ecosystem Assessment estimates that 35–50 percent of wetter, more fertile rangelands (e.g., temperate grasslands) have been converted to crop production, for example. They remain the most extensive type of land, however, encompassing some 40 percent of the ice-free terrain on the planet, roughly 1.5 times as much as the world's forests and 2.5 times as much as croplands.[4] Indeed, from an economist's perspective, rangelands suffer from excess supply: they are the least valuable territory, in money price per unit area, in nearly every society where they are found. There are many reasons for this, to be sure: steep terrain, rocky or infertile soils, low and erratic rainfall, extreme temperatures, and inaccessibility characterize many rangelands. But their biophysical marginality is alloyed with political and economic elements as well. Rangelands are perennial targets for development schemes of all sorts – factories, subdivisions, waste facilities, military installations, solar arrays, power plants, mines, you name it – if only because they proffer cheaper ground than other places. To paraphrase Marx, rangelands constitute a reserve acreage of under-employed lands at the ever-expanding fringes of the world's economic geography.

That rangelands appear "worthless" and "empty" is a powerful illusion, at once demonstrably false and perversely self-fulfilling. They support an estimated one billion people and supply animal protein, water, or other resources for twice that many. Their value for conservation is enormous, precisely because they are not plowed, paved over, or otherwise simplified by intensive human use. They hold approximately 30 percent of the world's soil carbon, and only tropical rainforests harbor greater biological diversity.[5] They may appear inhospitable, bewildering, or threatening from the vantage point of the "civilized" world of sedentary states and city-dwellers, but from the perspective of a nomadic pastoralist, rangelands are none of these things.[6] Yet like Marx's reserve army of the unemployed, they are rendered exploitable by the same forces that subsequently exploit them. The greatest threat to rangelands worldwide is land use change and fragmentation,[7] "made possible by two enabling conditions – the growing power of centralized, bureaucratic states and the spread of capitalism... the power of this combination is now felt even in remote, relatively unpopulated and economically marginal rangeland areas."[8] In the words of historian and geographer Diana Davis:

> The assumption that the world's drylands are worthless, deforested, and overgrazed landscapes has led, since the colonial period, to programs and policies that have often systematically damaged dryland environments and marginalized large numbers of indigenous peoples, many of whom had been using the land sustainably.[9]

To think about rangelands thus requires holding seemingly contradictory ideas side by side: at once diminished and vast, worthless and invaluable, marginal and pivotal. The narrative and conceptual space delineated by these polarities is thick with histories, both human and evolutionary, and also with parables, legends, stories, and speculations. The designation "Great American Desert," for example, which geographer Edwin James slapped onto the North American High Plains in his map for the Long Expedition in 1823, helped retard Euro-American settlement there for a half-century, not because it was true but because people took it to be true. Rangelands are enigmatic, liminal, beguiling spaces, and it can be difficult to distinguish fact from fiction, the actual from the imagined. Australian rangeland scientist Mark Stafford Smith has written that "In caricature, the relationship between centres of power and drylands falls into one of three categories – rape and pillage, well-intentioned but poorly understood intervention, or benign neglect."[10] The knowledge claims underlying Stafford Smith's second category have generally circulated under the sign of science, but rangelands have repeatedly induced a mix of wishful thinking, erasure, and hyperbole. Three ideas will serve to illustrate, two from the past and one that is making the rounds now.

Desertification

> Because deserts are perceived as the most "worthless" lands on the planet, the history of desertification and dryland development policies lay bare the political and economic foundations of our most common and influential desert imaginaries and our deeply capitalist relations with nature more generally.[11]
>
> Diana Davis

In late 1975, the Royal Swedish Academy of Sciences used its journal *Ambio* to publish an article entitled "Desertification: A World Problem," written by Erik Eckholm, a Senior Researcher at the Worldwatch Institute in Washington, DC. "Deserts are creeping outward in Africa, Asia, and Latin America," Eckholm began. This was not due to climate or drought, he argued, but to over-population and associated human impacts. "Populations are, in effect, outgrowing the biological systems that sustain their way of life... dessicated [sic], barren, desert-like lands are being *created*, a process that has become known as desertification." Overgrazing, fire, and imprudent cropping in "fragile" arid and semi-arid rangelands were altering the climate and reducing local rainfall in a vicious cycle. "It is a malignancy undermining the food-producing capacity of Africa, Asia, and Latin America," and it could only be stopped by urgent outside interventions to reduce livestock numbers, plant trees to slow erosion, and modernize production systems. "Human cultural patterns in the desert must be reshaped."[12]

Eckholm's piece was just one of many breathless declarations of a global desertification emergency in the early 1970s, a discourse enflamed by searing images of emaciated children and eviscerated livestock in the Sahelian region of West Africa. The hyperbolic tone suggested a novel, unprecedented threat, and indeed many people and countless livestock perished. But the narrative was more than a century old. Seemingly unbeknownst to the likes of Eckholm, desertification had its roots in nineteenth-century French colonial Morocco and Algeria, where scientific foresters diagnosed regional "desiccation" as the result of native herders' livestock and land management practices. As Diana Davis has shown, "The idea of desertification itself is in fact a colonial construction, a concept with little basis in empirical evidence initiated and propagated by those with a poor understanding of arid-land ecosystems."[13] Colonial administrators and professional scientists such as François Trottier, A.D. Combe, Paul Boudy and Charles Flahault, many of them trained at prestigious French universities, gave desertification the imprimatur of objective science. Whatever their individual motives may have been, their ideas "served three primary purposes: the appropriation of land and resources; social control (including the provision of labor); and the transformation of subsistence production into commodity production."[14] Similarly, the late-twentieth-century revival of the desertification narrative served powerful post-colonial interests in developing countries as well as the ascendant international development apparatus, giving rise in rapid order to a report from the U.S. Agency for International Development (1972), a UN General Assembly resolution (1974), an international conference and UN Plan of Action to Combat Desertification (1977), and eventually a permanent instrument, the UN Convention to Combat Desertification (1994).[15]

Several of the claims that gained the most traction in desertification discourse were quickly shown to be false (e.g., that the Sahara Desert was expanding southward at thirty miles per year); many others rested on incommensurable, spurious or non-existent data, dressed up in authoritative-sounding declarative prose (e.g., "at least 35 per cent of the earth's land surface is now threatened by desertification").[16] But banal factual refutations could not keep up with a narrative that, as in the colonial Maghreb, "was so useful to so many in positions of power who used it to justify their actions."[17] In the 1990s, arid lands expert Chuck Hutchinson pointed out that the strongest empirical case of desertification was the southwestern United States, but that went nowhere – demonstrating that the concept was really meant for use in poor countries.[18] Every decade or so, another compendium of scholarship documents the conceptual incoherence and empirical lacunae of desertification.[19] The latest of these calls Sahelian desertification "something that never occurred but was widely believed to have existed" and observes that the concept "has become a political tool of global importance even as the scientific basis for its use grows weaker."[20] Satellite remote sensing demonstrates unequivocally that the Sahel region has "re-greened" with better rainfall since the 1980s, and climate models now suggest that the severe drought of 1967–72 may actually have been driven by industrial aerosol emissions from Europe![21] But the desertification discourse marches on, as we will see, and the policy measures mobilized through the narrative – such as sedentarization of nomads, privatized land tenure, fencing, destocking to fixed carrying capacities, improved breeding, and agricultural intensification – continue to be advanced in many developing countries, despite repeated and well-documented failures.[22]

Succession

Frederic Clements is universally regarded as a major figure in the history of ecology, and his theory of plant succession is by all accounts the foundation on which the field of range science was built. Curiously, no authoritative biography of Clements has been written, and the existing literature about him rarely touches on the broader social and political contexts in which he lived and worked.[23] Born in Lincoln, Nebraska, in 1874, Clements grew up in the midst of one of the most dramatic episodes of landscape transformation in history: the breakneck conversion of the Great Plains from bison-dominated prairie to intensive grain agriculture in the span of a single generation. As a student of Charles Bessey and Roscoe Pound at the University of Nebraska, Clements fell under the spell of the prairie and its grasses, earning his doctorate at the age of 24 and immediately joining the faculty of botany. An ambitious scientist and obsessive worker, he was further motivated by the fact that the prairies around him were rapidly disappearing under the settlers' plows.

Nebraska provided Clements with an exceptional natural laboratory: a highly diverse but spatially continuous plant community that stretched 430 miles along a 15–35-inch west-to-east rainfall gradient. Using a meter-square quadrat method inspired by Bessey and perfected with Pound, Clements was able to document and measure the dynamic interactions of vegetation with rainfall and soils over space and time. Analyzing those dynamics with newly developed statistical techniques, he helped transform descriptive botany into modern ecology, grounded in the theory of plant succession. Largely on this basis, the University of Nebraska became the dominant force in American grassland ecology for decades to come, granting more than half of the nation's doctoral degrees in the field between 1895 and 1955 and training many of the men who invented range science, including Arthur Sampson, William Chapline, Jared Smith, and Clarence Forsling.[24]

Clements published his two-volume magnum opus, *Plant Succession* and *Plant Indicators,* in 1916 and 1920. The first volume developed his theory, and the second applied it (in encyclopedic detail) to the plant communities of the western United States. The very first paragraph of *Plant Indicators* explained its "practical aspect" and is worth quoting in full:

> Every plant is a measure of the conditions under which it grows. To this extent it is an index of soil and climate, and consequently an indicator of the behavior of other plants and of animals in the same spot. A vague recognition of the relation between plants and soil must have marked the very beginnings of agriculture. In a general way it has played its part in the colonization of new countries and the spread of cultivation into new areas, but the use of indicator plants in actual practice has remained slight. It is obviously of the greatest importance in newly settled regions. However, it is in just these regions that experience is lacking and correlation correspondingly difficult. In fact, the pioneer is often misled by his endeavor to transfer the experience gained in his former home to a new and different region. Differences of vegetation and climate, and often of soil as well, make a wholly new complex of relations. As a consequence, the settler is very apt to go astray in reaching conclusions as to the significance of a particular plant. As the country becomes more settled, experience accumulates and makes it increasingly possible to recognize helpful correlations. But this period usually passes too quickly to establish a procedure before the native plants have disappeared, except from roadsides, meadows, and pastures. The manner and degree of utilization of natural meadows and pastures are clearly indicated by the plants in them. Yet it is exceptional that these indicators are recognized and made use of by the farmer.[25]

A successional understanding of native vegetation could help identify the cultivars to which any given site was suited by its soils and climate, facilitating rapid and efficient installation of commercially viable farms. This theme recurs throughout the book, as Clements remarks in passing on costly – or even tragic – mistakes in planting choices by settlers in various locations. The express intent of Clements's theory, then, was to aid in successful colonization and agricultural settlement in "new countries" – an anodyne allusion to the then-still-recent conquest and dispossession of Native Americans. Stronger confirmation of Libby Robin's contention that ecology is a "science of empire" could scarcely be imagined.[26]

Much of the western U.S. was rangeland, and Clements paid particular attention to how succession could aid in grazing management. Droughts and winter storms during the Cattle Boom of 1873–93 had resulted in massive livestock die-offs and widespread, persistent rangeland degradation; how to remedy this damage while still supporting settlement was an urgent question. Clements's theory provided a reassuringly positive answer, grounded in an analogy between cattle and bison. Reviewing the accounts of early explorers and migrants, Clements found wide disparities: some described endless expanses of grass as tall as a horse, others a landscape nearly denuded of vegetation. Both were accurate, he averred: they just happened to witness different moments in the dynamic interplay of bison, rainfall, fire, and grasses. "All the statements agree as to the excessive damage done to the range by buffalo, but it seems certain that the more or less complete rest which followed brought about a fair degree of recovery in a few years."[27] Likewise with fire or drought: the bison would simply migrate elsewhere, allowing the disturbed areas to recover.

> It is obvious that an area destructively overgrazed would be abandoned by grazing animals for an untouched portion of the same climax, and that the bare area would then pass through the various stages of succession to again reach the climax in 20 to 30 years.[28]

The core ideas of Clements's theory were elegantly demonstrated in this simple, archetypal case: following disturbance, a plant community passes through stages of recovery (succession) until it returns to its equilibrium state (climax), provided the disturbing agent is removed. Cattle might not be able to migrate long distances, but their owners could simulate the process by rotating their herds between multiple pastures.

> The recognition of past and present cycles of overgrazing is of great practical importance. Its greatest value lies in the certainty that a range will return to its normal condition once it is given a chance to regenerate... all overgrazed ranges can be certainly and greatly improved by proper rest or rotation. This is the basis of all range improvement.[29]

Clements was explicit about both ends and means: "The primary object of range improvement is to secure and maintain the maximum carrying capacity. The chief factors in this are proper stocking and rotation grazing."[30] The result was managerial control: "an elementary understanding of successional processes furnishes a tool for manipulating the grazing cover more or less as desired."[31]

Clements believed that his theory of plant succession was "of universal application," valid not only in Nebraska or the Great Plains but throughout the world, forwards and backwards in time even on geological timescales.[32] This was crucial to claiming the mantle of a rigorous, formal science at the time, and it certainly aided the adoption of successional theory as the basis for range management in the U.S. Forest Service, which dominated range science (as well as forestry) through the first half of the twentieth century.[33] But it was also an extravagant over-generalization. As Ronald Tobey points out, the droughts of the 1930s led to a wholesale shift in the composition of the prairies in eastern Nebraska even during Clements's lifetime, much to the alarm and dismay of his famous disciple and collaborator, John Weaver.[34] Degraded grassland sites in the southwestern U.S. where livestock were removed early in the twentieth century failed to conform to successional expectations, instead converting to shrub dominance. But alternative paradigms did not emerge until the 1970s and 1980s.

Only in the present century have scientists replaced Clementsian succession altogether for sites where the coefficient of variation of inter-annual precipitation exceeds 33 percent – which is not the case in Nebraska, but is true for roughly 28 percent of the world's rangelands.[35] These are now understood as non-equilibrium systems, with complex, non-linear dynamics and multiple stable states, in which abiotic factors (e.g., drought, rainfall, fire or frost) are often the main drivers of change. Recognition of this has forced a fundamental rethinking of "degradation," which had previously been defined as departure-from-climax (or various analogues thereof). Thus, rangelands at the drier, more variable end of the spectrum – such as the Sahel – are now considered *less* fragile than before, and globally less degraded than more temperate grasslands.[36] The concept of carrying capacity as a singular, static attribute of rangelands has been widely debunked, as "average" forage production almost never obtains and livestock–vegetation dynamics are simply too varied and complex to be captured by such a blunt instrument.[37] The role of fire has likewise been re-evaluated: rather than an unmitigated evil, as it was deemed by French colonial and U.S. Forest Service officials alike, fire is now seen as an unavoidable and often beneficial ecological process on many rangelands.

In short, the Euro-American conventional wisdom about rangelands that prevailed from the early nineteenth to the late twentieth centuries has been upended, at least among scholars and scientists. In addition to new models for how rangelands function, there is growing recognition that long-time rangeland inhabitants and managers – including pastoralists, indigenous groups, and multi-generation ranchers – possess important local and traditional ecological knowledge that has heretofore been overlooked or actively dismissed by professional experts. This is not to say, however, that the broader public or policymakers have absorbed the new findings, nor that scholars and scientists today are necessarily immune to the mistakes that afflicted their predecessors.

Rangelands as Climate Solution

Combatting anthropogenic climate change is a common theme of discussions about rangelands in the twenty-first century. In various combinations, people from the environmental NGO community, ranching, philanthropy, academic science, government, and multilateral agencies have rallied around this cause, employing partially overlapping vocabularies to explain and energize their efforts: holistic management, planned grazing, regenerative agriculture, or adaptive, multi-paddock (AMP) grazing, for example. They should not be conflated, if only because they often see themselves as quite distinct, but a full examination of their differences is beyond my scope here. It is fair to say, nonetheless, that they have at least three things in common: a genuine and growing concern about climate change; a belief that rangelands, and specifically rangeland soils, can play a major role in addressing the problem; and a commitment to rotational grazing as a primary management practice for achieving this vision. What light can the history described above shed on these ideas, and vice versa?

Allan Savory's 2013 TED Talk, "How to Fight Desertification and Reverse Climate Change," has been viewed more than 7.75 million times. Described on the TED website as a "grassland ecosystem pioneer," Savory is the charismatic and controversial founder of holistic management, with ardent supporters and detractors alike. In his talk, he asserts that "about two-thirds… of the world is desertifying," and that land degradation is as important (and "maybe" more important) than fossil fuels as a driver of climate change. His core claims are captured in these passages:

> There is only one option, I repeat to you, only one option left to climatologists and scientists, and that is to do the unthinkable, and to use livestock, bunched and moving, as a proxy for former herds and predators, and mimic nature. There is no other alternative left to mankind… [I]f we do what I am showing you here, we can take enough carbon out of the atmosphere and safely store it in the grassland soils for thousands of years, and if we just do that on about half the world's grasslands that I've shown you, we can take us back to pre-industrial levels while feeding people.[38]

Savory does not mention bison or the Great Plains, and he does not acknowledge that his model and inspiration – large mobile herds of wildlife – is identical to that of Clements a century earlier, nor that rotational grazing, too, is a venerable subject in the literature, as we have already seen. Roy Behnke's summation is incisive:

> Between 1948 and 2003 roughly two out of every five articles in the *Journal of Range Management*… were about fenced "rotational" grazing systems… Despite the decades of negative or mixed results [from research], the debate about the efficacy of rotational systems in semi-arid rangelands grinds on without resolution. The safest conclusion may be that the advantages of rotational systems are

either modest and difficult to detect, or so contingent upon local circumstances or skilled management as to make them difficult to replicate. Irrespective of the ultimate outcome of the debate, at this late date rotational grazing seems unlikely to produce any dramatic breakthroughs.[39]

Savory's claims are extreme, but his underlying arguments bear a strong resemblance to those advanced less hyperbolically in other circles. Proponents of regenerative agriculture and AMP, for example, also emphasize the potential of rotational grazing to sequester carbon in rangeland soils and thereby mitigate climate change.[40] The application of compost or other organic fertilizers to rangelands has also been studied and advocated as a climate mitigation strategy because of its potential to augment soil carbon stocks.[41]

Implicitly or expressly, the vast extent of the world's rangelands is a key plank in all these platforms. With approximately 3.4 billion hectares to work with, even small net gains in soil carbon per hectare could make very large contributions to combatting climate change. But the problem of over-extrapolation here is manifold. There is the practical matter of scaling up: Most study sites are <40 hectares in size, after all. Savory claims that 15 million hectares of land are already engaged in his effort, but the total he is invoking for his "solution" is minimally about 50 times that much, and taking him literally would suggest closer to 200 times that amount.[42] Moreover, for all of these proposals, even if practical obstacles could be surmounted, there is no reason to believe that results would be consistent across the world's diverse soils, climates, vegetation types, baseline conditions, and livestock systems. The most comprehensive meta-analyses find potential for enhanced soil carbon sequestration in rangelands, but with very significant caveats: heavy stocking may lead to net soil carbon losses; net effects may be driven not by management but by abiotic factors, especially in more xeric rangelands; soil carbon increases may be short-lived and/or intractable to reliable measurement. As one study cautions,

> these results do not apply uniformly to all grazing lands and extrapolating the results of this synthesis regionally or globally requires information about where there is scope for improvement of grassland management… it is not always the case that improved grazing management leads to increased soil C stocks. Even when it does, soil C stock responses vary as a function of climate, soil, and vegetation characteristics.[43]

On top of these issues – which can be extremely technical – is a simpler and arguably more decisive weakness: land *use* swamps management in any large-scale assessment of rangeland soil carbon. Many studies indicate that retiring croplands is the most effective technique, per hectare, for carbon sequestration, and there is also strong evidence that avoiding the conversion of existing rangelands to other uses may be the highest priority for net carbon sequestration overall.

Conclusion

If there is any one theme that emerges from the history of rangelands since 1492, it is recursive misapprehension: seemingly authoritative or "expert" knowledge about rangelands has repeatedly turned out to be exaggerated, shortsighted, incomplete, or just plain wrong. This is not simply a matter of science and reason incrementally overcoming ignorance, moreover. I have previously written that

> Rangelands are sites where the separate and combined efforts of capital, science and the state meet their limits, not in any fixed sense but as part of ongoing processes of trying to overcome and extend those limits... It is precisely their manifold marginality that enables rangelands to defy and disrupt social forces that elsewhere seem so powerful, and thereby to illuminate core tendencies, contradictions, and limitations in modern ways of knowing, using, and governing lands and people.[44]

In a handful of individual cases, such as the Coyote-Proof Pasture Experiment, historical analysis can demonstrate conclusively that the "scientific" basis of rangeland policies was pre-determined and manipulated to suit political-administrative exigencies.[45] But it is generally unwise to speculate on the inner thoughts and motives of people, and whether scientists such as Boudy, Flahault, Clements, and Eckholm intended (or even recognized) their complicity in colonial and postcolonial injustices is impossible to judge. It is fair to say, however, that their knowledge claims reflected large measures of wishful thinking, erasure, and hyperbole. Wittingly or not, they discerned the evidence that suited their (or their patrons') needs and expectations, while downplaying, overlooking, or omitting evidence and arguments that did not. To call their conclusions false is not really much better than calling them true, however: either judgment would exaggerate the degree of certainty available, then or now. As Mark Stafford Smith remarks, desertification is a mirage: "But of course mirages are real phenomena, just not the ones that they appear to be... And if you march across the desert through one mirage, as often as not another appears ahead of you."[46]

The mirages that afflict thinking about rangelands are induced in part by the word itself. Even just employing the term attributes a measure of unity and coherence to what are in fact infinitely varied and diverse landscapes, while erasing the negative and residual nature of the category as well as the colonial and neo-colonial positionalities embedded in its construction. Like "nature" and "the state," then, rangelands should be handled with great epistemological care – or what Pierre Bourdieu describes as *hyperbolic doubt*, questioning "all the presuppositions inscribed in the reality under analysis as well as in the very thoughts of the analyst."[47] The propensity to find in rangelands a solution to climate change, for example, must be interrogated not only in relation to the relevant scientific literature, but also in light of the political and economic marginality of rangelands relative to other land types and other land uses. Surely the fact that rangelands are less expensive than other lands – more abundant and also less politically powerful – conditions the enthusiasm of proponents of holistic management, regenerative agriculture, AMP grazing and compost application? Retiring agricultural lands would be a more reliable and effective way to sequester carbon in soils, per hectare, but it would also be much more costly and contentious; prohibiting the conversion of rangelands to other uses would eliminate capitalism's reserve of "under-employed" land and provoke pitched battles over private property rights. And, of course, the real solution to climate change – rapidly phasing out the combustion of fossil fuels – still strikes many people as impossibly utopian. Compared to all of these ideas, multiplying the results of a handful of controlled experiments by the world's 3.4 billion hectares of rangelands is tantalizingly easy. But pinning humanity's hopes on rangelands recapitulates a long history of wishful thinking, erasure and hyperbole perpetrated by outsiders – meaning non-rangeland residents – that is littered with policy failures and injustices.

Stewardship of rangelands in the twenty-first century must contend not only with the complexity of the landscapes in question but also with the political-economic forces that relegate them to the margins of power. This will require new narratives that elevate rangelands for their beauty and positive values, rather than just their vast extent and putative degradation. To succeed, it will also require strategies that engage and strengthen local communities and institutions vis-à-vis outside forces – including scientists – whose ambitions, when not openly predatory, are often still suffused with flawed assumptions and wishful thinking.

Notes

1 Alfred W. Crosby, *Ecological Imperialism: The Biological Expansion of Europe, 900–1900*, 2nd edition (Cambridge: Cambridge University Press, 2004), 196.
2 Richard White, "Animals and Enterprise," in *The Oxford History of the American West*, eds. Clyde A. Milner II, Carol A. O'Connor and Martha A. Sandweiss (New York, NY: Oxford University Press, 1994), 238.
3 Anthony C. Grice and Kenneth C. Hodgkinson, *Global Rangelands: Progress and Prospects* (Wallingford, UK: CABI Publishing, 2002), 2.

4 Gregory P. Asner, Andrew J. Elmore, Lydia P. Olander, Roberta E. Martin and A. Thomas Harris, "Grazing Systems, Ecosystem Responses, and Global Change," *Annual Review of Environmental Resources* 29 (2004), 261–99.
5 Millennium Ecosystem Assessment synthesis reports, 2005. http://www.millenniumassessment.org/en/Synthesis.html, viewed 21 September 2021.
6 The deep history of nomadic pastoralists as "barbarians" in relation to states of all kinds parallels present-day dynamics in important ways. "Barbarian geography and ecology… constitutes a large and residual category; basically they comprise all those geographies that are unsuitable for state making." James C. Scott, *Against the Grain: A Deep History of the Earliest States* (New Haven CT: Yale University Press, 2017), 228.
7 N. Thompson Hobbs, Robin S. Reid, Kathleen A. Galvin, and James E. Ellis, "Fragmentation of Arid and Semi-arid Ecosystems: Implications for People and Animals," in *Fragmentation in Semi-Arid and Arid Landscapes: Consequences for Human and Natural Systems*, eds. Roy H. Behnke, Jr., Kathleen A. Galvin, N. Thompson Hobbs and Robin S. Reid (Dordrecht: Springer, 2008), 25–44.
8 Roy H. Behnke, "The Drivers of Fragmentation in Arid and Semi-Arid Landscapes," 305–340 in *Fragmentation in Semi-Arid and Arid Landscapes: Consequences for Human and Natural Systems*, eds. Roy H. Behnke Jr., Kathleen A. Galvin, N. Thompson Hobbs and Robin S. Reid (Dordrecht: Springer, 2008), 306.
9 Diana K. Davis, *The Arid Lands: History, Power, Knowledge* (Cambridge, MA: The MIT Press, 2016), 4.
10 Mark Stafford Smith, "Desertification: Reflections on the Mirage," 539–60 in *The End of Desertification? Disputing Environmental Change in the Drylands*, eds. Roy H. Behnke and Michael Mortimore (Heidelberg: Springer, 2016), 554.
11 Davis, *The Arid Lands*, 168.
12 Erik P. Eckholm, "Desertification: A World Problem," *Ambio* 4, no. 4 (1975): 137–145, 138, 140 (emphasis in original).
13 Davis, *Resurrecting the Granary of Rome: Environmental History and French Colonial Expansion in North Africa* (Athens: Ohio University Press, 2007), 171.
14 Diana K. Davis, *Resurrecting the Granary of Rome*, 165–66.
15 Nathan F. Sayre, *The Politics of Scale: A History of Rangeland Science* (Chicago: University of Chicago Press, 2017), 181.
16 David S.G. Thomas and Nick Middleton, *Desertification: Exploding the Myth* (Chichester, UK: Wiley, 1994), 52.
17 Davis, *The Arid Lands*, 95.
18 Charles F. Hutchinson, "The Sahelian Desertification Debate: A View from the American South-west," *Journal of Arid Environments* 33 (1996), 519–524.
19 For the 1990s, see Thomas and Middleton, *Desertification*. For the 2000s, see James F. Reynolds and Mark Stafford Smith, eds., *Global Desertification: Do Humans Cause Deserts?* (Berlin: Dahlem University Press, 2002). For the 2010s, see Behnke and Mortimore, *The End of Desertification?*
20 Behnke and Mortimore, *The End of Desertification?*, 1, 2.
21 Cecile Dardel, Laurent Kergoat, Pierre Hiernaux, Olivier Mougin, Michel Grippa and C.J. Tucker, "Re-greening Sahel: 30 Years of Remote Sensing Data and Field Observations (Mali, Niger)," *Remote Sensing of Environment* 140 (January 2014), 350–64; Michela Biasutti, "A Man-made Drought." *Nature Climate Change* 1 (July 2011): 197–98.

22 The Chinese government's policies in Inner Mongolia and Tibet are among the most disturbing examples of recent years. See, e.g., Wenjun Li and Lynn Huntsinger, "China's Grassland Contract Policy and its Impacts on Herder Ability to Benefit in Inner Mongolia: Tragic Feedbacks," *Ecology and Society* 16 (June 2011): 1.
23 The closest thing to a biography that exists was penned by his wife and scientific collaborator, Edith, under the title *Adventures in Ecology: Half a Million Miles: From Mud to Macadam* (New York: Pageant Press, 1960). Ronald C. Tobey's *Saving the Prairies: The Life Cycle of the Founding School of American Plant Ecology, 1895–1955* (Berkeley: University of California Press, 1981) also contains important biographical materials. For an article-length summary of Clements's influence in range science, see Linda A. Joyce, "The Life Cycle of the Range Condition Concept," *Journal of Range Management* 46 (March 1993): 132–38. See also Sayre, *The Politics of Scale*.
24 Tobey, *Saving the Prairies*, 120; Sayre, *The Politics of Scale*.
25 Frederic E. Clements, *Plant Indicators: The Relation of Plant Communities to Process and Practice* (Washington: Carnegie Institution of Washington, 1920), 3.
26 Libby Robin, "Ecology: A Science of Empire?," in Tom Griffiths and Libby Robin, eds., *Ecology and Empire: Environmental History of Settler Societies* (Seattle: University of Washington Press, 1997), 63–75.
27 Clements, *Plant Indicators*, 310.
28 Clements, *Plant Indicators*, 307. The last section (20 to 30 years) was little more than a guess, as Clements had only been observing grasses for that long himself and the southern bison herd had been hunted to oblivion in the three years prior to his birth.
29 Clements, *Plant Indicators*, 310.
30 Clements, *Plant Indicators*, 312.
31 Clements, *Plant Indicators*, 311.
32 Frederic E. Clements, *Plant Succession: An Analysis of the Development of Vegetation* (Washington: Carnegie Institution of Washington, 1916), iii.
33 Earle Clapp, head of research for the Forest Service from 1915 to 1935, embraced succession as the basis of range research even as he acknowledged that "Practically the entire question of association development and succession… still awaits investigation." Sayre, *The Politics of Scale*, 69.
34 Tobey, *Saving the Prairies*, 201.
35 Henrik von Wehrden, Jan Hanspach, Petra Kaczensky, Joern Fischer and Karsten Wesche, "Global Assessment of the Non-equilibrium Concept in Rangelands," *Ecological Applications* 22 (2012), 393–99; Nathan F. Sayre, Diana K. Davis, Brandon Bestelmeyer and Jeb C. Williamson, "Rangelands: Where Anthromes Meet Their Limits," *Land* 6 (2017): 31.
36 According to a 2009 review by the UN's Food and Agriculture Organization, 22 percent of the world's drylands are considered degraded, compared to up to 71 percent of the world's grasslands. Constance Neely, Sally Bunning and Andreas Wilkes, eds., *Review of Evidence on Drylands Pastoral Systems and Climate Change: Implications and Opportunities for Mitigation and Adaptation* (Rome: Food and Agriculture Organization of the United Nations, 2009).
37 Clements pointed this out multiple times in *Plant Indicators*, but for various reasons the point was lost in both policy and practice (see Sayre, *The Politics of Scale*). For an influential critique of carrying capacity, see Roy H. Behnke, Jr., Ian Scoones and Carol Kerven, eds., *Range Ecology at Disequilibrium: New Models of Natural Variability and Pastoral Adaptation in African Savannas* (London: Overseas Development Institute, 1993).

38 https://www.ted.com/talks/allan_savory_how_to_fight_desertification_and_reverse_climate_change?language=en.
39 Roy H. Behnke, "Grazing Into the Anthropocene *or* Back to the Future?," *Frontiers in Sustainable Food Systems* 5 (2021), article 638806, 5.
40 Paige L. Stanley, Jason E. Rowntree, David K. Beede, Marcia S. DeLonge and Michael W. Hamm, "Impacts of Soil Carbon Sequestration on Life Cycle Greenhouse Gas Emissions in Midwestern USA Beef Finishing Systems," *Agricultural Systems* 162 (2018): 249–58; Rebecca Ryals, Melannie D. Hartman, William J. Parton, Marcia S. DeLonge and Whendee L. Silver, "Long-term Climate Change Mitigation Potential with Organic Matter Management on Grasslands," *Ecological Applications* 25 (2015): 531–45.
41 Rebecca Ryals and Whendee L. Silver, "Effects of Organic Matter Amendments on Net Primary Productivity and Greenhouse Gas Emissions in Annual Grasslands," *Ecological Applications* 23 (January 2013): 46–59.
42 Half of the world's grasslands would equal some 711 million hectares; half of the savannas, grasslands, deserts, and open shrublands would equal slightly over 3 billion hectares – which is more than the area of those biomes that is currently grazed by livestock. See Asner et al., "Grazing Systems, Ecosystem Responses, and Global Change."
43 Richard T. Conant, Carlos E.P. Cerri, Brooke B. Osbourne and Keith Paustian, "Grassland Management Impacts on Soil Carbon Stocks: a New Synthesis." *Ecological Applications* 27 (2017): 662–8, 666.
44 Sayre, *The Politics of Scale*, 2.
45 Sayre, *The Politics of Scale*, chapter 1.
46 Stafford Smith, "Desertification: Reflections on the Mirage," 540.
47 Pierre Bourdieu, "Rethinking the State: Genesis and Structure of the Bureaucratic Field," *Sociological Theory* 12 (March 1994): 1–18, 1.

Bibliography

Asner, Gregory P., Andrew J. Elmore, Lydia P. Olander, Roberta E. Martin and A. Thomas Harris. "Grazing Systems, Ecosystem Responses, and Global Change." *Annual Review of Environmental Resources* 29 (2004): 261–299.

Behnke, Roy H. "The Drivers of Fragmentation in Arid and Semi-Arid Landscapes." In *Fragmentation in Semi-Arid and Arid Landscapes: Consequences for Human and Natural Systems*, edited by Roy H. Behnke, Jr., Kathleen A. Galvin, N. Thompson Hobbs and Robin S. Reid, 305–340. Dordrecht: Springer, 2008.

Behnke, Roy H. "Grazing into the Anthropocene *or* Back to the Future?" *Frontiers in Sustainable Food Systems* 5 (2021), article 638806.

Behnke, Roy H., Jr., Ian Scoones and Carol Kerven, eds. *Range Ecology at Disequilibrium: New Models of Natural Variability and Pastoral Adaptation in African Savannas*. London: Overseas Development Institute, 1993.

Biasutti, Michela. "A Man-made Drought." *Nature Climate Change* 1 (July 2011): 197–198.

Bourdieu, Pierre. "Rethinking the State: Genesis and Structure of the Bureaucratic Field." *Sociological Theory* 12 (March 1994): 1–18.

Clements, Edith. *Adventures in Ecology; Half a Million Miles: From Mud to Macadam*. New York, NY: Pageant Press, 1960.

Clements, Frederic E. *Plant Indicators: The Relation of Plant Communities to Process and Practice*. Washington, DC: Carnegie Institution of Washington, 1920.

Clements, Frederic E. *Plant Succession: An Analysis of the Development of Vegetation.* Washington, DC: Carnegie Institution of Washington, 1916.

Conant, Richard T., Carlos E.P. Cerri, Brooke B. Osbourne and Keith Paustian. "Grassland Management Impacts on Soil Carbon Stocks: A New Synthesis." *Ecological Applications* 27 (2017): 662–668.

Crosby, Alfred W. *Ecological Imperialism: The Biological Expansion of Europe, 900–1900*, 2nd edition. Cambridge: Cambridge University Press, 2004.

Dardel, Cecile, Laurent Kergoat, Pierre Hiernaux, Olivier Mougin, Michel Grippa and C.J. Tucker. "Re-greening Sahel: 30 Years of Remote Sensing Data and Field Observations (Mali, Niger)." *Remote Sensing of Environment* 140 (January 2014): 350–364.

Davis, Diana K. *The Arid Lands: History, Power, Knowledge.* Cambridge, MA: The MIT Press, 2016.

Davis, Diana K. *Resurrecting the Granary of Rome: Environmental History and French Colonial Expansion in North Africa.* Athens, OH: Ohio University Press, 2007.

Eckholm, Erik P. "Desertification: A World Problem." *Ambio* 4, no. 4 (1975): 137–145.

Grice, A.C. and K.C. Hodgkinson. *Global Rangelands: Progress and Prospects.* Wallingford, UK: CABI Publishing, 2002.

Hobbs, N. Thompson, Robin S. Reid, Kathleen A. Galvin, and James E. Ellis. "Fragmentation of Arid and Semi-arid Ecosystems: Implications for People and Animals." In *Fragmentation in Semi-Arid and Arid Landscapes: Consequences for Human and Natural Systems*, edited by Roy H. Behnke, Jr., Kathleen A. Galvin, N. Thompson Hobbs and Robin S. Reid, 25–44. Dordrecht: Springer, 2008.

Hutchinson, Charles F. "The Sahelian Desertification Debate: A View from the American South-west." *Journal of Arid Environments* 33 (1996): 519–524.

Joyce, Linda A. "The Life Cycle of the Range Condition Concept." *Journal of Range Management* 46 (March 1993): 132–138.

Li, Wenjun and Lynn Huntsinger. "China's Grassland Contract Policy and its Impacts on Herder Ability to Benefit in Inner Mongolia: Tragic Feedbacks." *Ecology and Society* 16, no. 2 (June 2011), article 1.

Millennium Ecosystem Assessment Synthesis Reports, 2005. http://www.millenniumassessment.org/en/Synthesis.html, viewed September 21, 2021.

Neely, Constance, Sally Bunning and Andreas Wilkes, eds. *Review of Evidence on Drylands Pastoral Systems and Climate Change: Implications and Opportunities for Mitigation and Adaptation.* Rome: Food and Agriculture Organization of the United Nations, 2009.

Reynolds, James F. and Mark Stafford Smith, eds. *Global Desertification: Do Humans Cause Deserts?* Berlin: Dahlem University Press, 2002.

Robin, Libby. "Ecology: A Science of Empire?" In *Ecology and Empire: Environmental History of Settler Societies*, edited by R. Griffiths and L. Robin, 63–75. Seattle: University of Washington Press, 1997.

Ryals, Rebecca, Melannie D. Hartman, William J. Parton, Marcia S. DeLonge and Whendee L. Silver. "Long-term Climate Change Mitigation Potential with Organic Matter Management on Grasslands." *Ecological Applications* 25 (2015): 531–545.

Ryals, Rebecca and Whendee L. Silver. "Effects of Organic Matter Amendments on Net Primary Productivity and Greenhouse Gas Emissions in Annual Grasslands." *Ecological Applications* 23 (January 2013): 46–59.

Savory, Allan. "How to Fight Desertification and Reverse Climate Change." *TED Talk*, 2013. https://www.ted.com/talks/allan_savory_how_to_fight_desertification_and_reverse_climate_change?language=en.

Sayre, Nathan F. *The Politics of Scale: A History of Rangeland Science*. Chicago, IL: University of Chicago Press, 2017.

Sayre, Nathan F., Diana K. Davis, Brandon Bestelmeyer and Jeb C. Williamson. "Rangelands: Where Anthromes Meet Their Limits." *Land* 6, no. 2 (2017): 1–11.

Scott, James C. *Against the Grain: A Deep History of the Earliest States*. New Haven, CT: Yale University Press, 2017.

Stafford Smith, Mark. "Desertification: Reflections on the Mirage." In *The End of Desertification? Disputing Environmental Change in the Drylands*, edited by Roy H. Behnke and Michael Mortimore, 539–560. Heidelberg: Springer, 2016.

Stanley, Paige L., Jason E. Rowntree, David K. Beede, Marcia S. DeLonge and Michael W. Hamm. "Impacts of Soil Carbon Sequestration on Life Cycle Greenhouse Gas Emissions in Midwestern USA Beef Finishing Systems." *Agricultural Systems* 162 (2018): 249–258.

Thomas, David S.G. and Nick Middleton. *Desertification: Exploding the Myth*. Chichester, UK: Wiley, 1994.

Tobey, Ronald C. *Saving the Prairies: The Life Cycle of the Founding School of American Plant Ecology, 1895–1955*. Berkeley, CA: University of California Press, 1981.

von Wehrden, Henrik, Jan Hanspach, Petra Kaczensky, Joern Fischer and Karsten Wesche. "Global Assessment of the Non-equilibrium Concept in Rangelands." *Ecological Applications* 22 (2012): 393–399.

White, Richard. "Animals and Enterprise." In *The Oxford History of the American West*, edited by Clyde A. Milner, II, Carol A. O'Connor and Martha A. Sandweiss, 237–273. New York, NY: Oxford University Press, 1994.

Part IV
Envisioning the Future

10 Product Stewardship

Ethics and Effectiveness in a Circular Economy

Helen Lewis and Nick Florin

Product stewardship suggests a pathway to more sustainable production, use, and disposal of manufactured products and packaging. The term is used to describe a wide variety of approaches that allocate responsibility for the sustainable management of products, particularly at end of life.[1]

The idea of product stewardship is highly contested, however, with different perspectives on what needs to happen to reduce the environmental and social impacts of products and who should be held responsible. Conventional ideas of product stewardship are also challenged by recent circular economy discourses, which propose more innovative solutions to the consumption of non-renewable resources, growing levels of waste and pollution, and the adverse impacts of complex global supply chains on human and environmental health.

This chapter explores historical product stewardship discourses from two perspectives: the values or principles that underpin them, and their relevance to a circular economy in the twenty-first century. Considering the latter, we discuss some limitations of product stewardship in a circular economy and we also reflect on the significant potential for product stewardship to support a deep transition to a circular economy that might address fundamental societal challenges linked to unsustainable rates of consumption.

Social Meanings of Product Stewardship

The term product stewardship means different things to different people depending on their values and the context within which it is being used.

Two key variables are the extent to which a scheme or program is regulated, and the degree of producer responsibility compared to other sectors in the value chain as well as consumers (Figure 10.1). Here responsibility is considered for the whole extent of a product's life cycle and this may be physical responsibility and/or economic responsibility. While there are many variations in terminology, policies and programs that place mandatory obligations on to producers, particularly at end of life, are often referred to as extended producer responsibility (EPR).

DOI: 10.4324/9781003219064-15

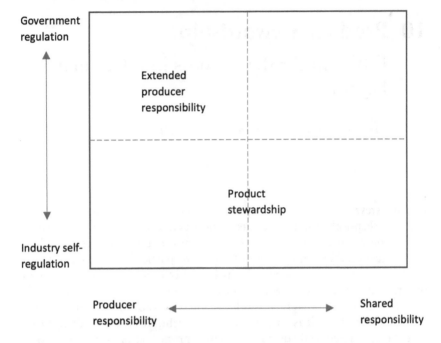

Figure 10.1 Key differences between EPR and product stewardship[2]

Product stewardship is often used to refer to policies or programs that involve self-regulation by liable firms, and some degree of shared responsibility between different sectors in the value chain. Social meanings attached to product stewardship include its importance as a risk management tool, particularly in the chemicals industry, and as a source of shared value creation. This section describes some of the more common interpretations and what they may mean in practice.

Extended Producer Responsibility for Waste

As discussed above, product stewardship is commonly associated with the principle of EPR. The Organisation for Economic Co-operation and Development (OECD) define EPR as "an environmental policy approach in which a producer's responsibility for a product is extended to the post-consumer stage of a product's life cycle."[3] This goes beyond the traditional responsibilities of a producer or importer, such as worker safety, pollution from manufacturing processes, and the sound management of production wastes. EPR emerged in the late 1980s in response to the challenges being faced by municipalities in managing growing volumes and increasing complexity of product wastes.

Product stewardship is often used to describe voluntary initiatives to address impacts at any stage of the product life cycle. In contrast, EPR is generally enforced through regulations that impose a financial or physical responsibility on producers. The policy rationale is that by shifting the financial burden of waste management to producers (manufacturers and importers) they will have an incentive to design products that can be recycled or safely disposed at end of life. EPR therefore operates as a market-based policy instrument to implement the "polluter pays principle," although in this context the polluter is regarded as the economic agent that has the greatest ability to avoid pollution through changes in design.[4]

In 2013 the OECD estimated that around 395 EPR policies were operating around the world. This was based on a fairly wide interpretation of EPR policies, including take-back schemes, advance disposal fees, deposit/refund systems, upstream taxes or subsidies, and virgin material taxes.[5]

A typical example of EPR being applied to waste management is the European Battery Directive (2006/66/EC) that sets out an approach for collecting and recycling batteries at end of life.[6] The broad goal of the policy is to minimise the negative environmental impacts of batteries. The Directive explicitly prohibits placing products on the market that contain mercury, cadmium, and other hazardous substances; it establishes minimum targets for the rate of collection of batteries at end of life; and it defines a benchmark efficiency for recycling processes by setting a minimum mass of material to be recovered. Aligned with EPR principles, the Directive provides guidance on the financial liabilities of producers whereby the collection of batteries is financed by the producers. It is stipulated that producers of all portable batteries set up convenient and free channels for collection, and all producers of industrial and automotive batteries are required to take back all waste batteries with no charge to end users. Approaches to implement the Directive and the outcomes vary owing to differences in how it is enforced through regulation developed by the member states.

Product Stewardship as Shared Responsibility

Another common meaning given to product stewardship is that of "shared responsibility" for the life cycle management of products. This interpretation allocates responsibility more broadly than EPR to include all of the organisations, as well as consumers, that can influence the impacts of a product over its full life cycle. It recognises that producers have most control over how products are designed and the choice of materials, but less influence on infrastructures for collection and recycling, or the behaviour of consumers.

The Australian Packaging Covenant Organisation (APCO) is an example of a product stewardship organisation (PSO) that administers an industry-led program with broad membership spanning the whole packaging value chain, including importers, manufacturers, and retailers. The Australian Packaging Covenant is established based on a "co-regulatory" model, which means it is led by industry and supported by government regulation to minimise "free-riding" organisations that may benefit from the program without contributing. APCO makes this clear on their website:

> To ensure our goals are met – APCO, its Members, industry, and state and federal governments all agree to comply with the obligations set out within the Covenant document. This co-regulatory framework recognises that all sectors and governments have a role to play, working together to find the best possible solutions for packaging efficiency and sustainability in Australia.[7]

Product Stewardship as a Risk Management Tool

Originally, product stewardship had a very different meaning and purpose. It first appeared in a voluntary code of practice developed by the Canadian Chemical Producers' Association (CCPA) in 1981 called Responsible Care/Product Stewardship.[8] In November 1984 a chemical storage plant owned by Union Carbide in Bhopal, India exploded, killing around 7,000 people in the first three days and many thousands of people since.[9] After this event the CCPA Board decided to develop Responsible Care into a proactive safety audit process with an emphasis on product stewardship. There were concerns within the chemical industry that unless they took strong action, they would lose their social license to operate.[10]

Responsible Care continues to be implemented in more than 65 countries. It includes the product stewardship management code

> for assuring the safe handling and use of chemicals, throughout each chemical's life cycle, that is from R&D and design through to manufacturing, marketing, distribution, use, recycling and disposal of chemical products.... Within this code, risk management principles are regarded as the cornerstone of Product Stewardship.[11]

In its historical context product stewardship was applied to bulk chemicals – some as ingredients in products – rather than individual items purchased by consumers.

Product Stewardship as Shared Value Creation

"Shared value" is the principle that companies should act in ways that create value for their business as well as society.[12] In a comprehensive review of sustainable business models, Bocken et al. categorized eight sustainable business model archetypes, including "adopt a stewardship role."[13] This categorization recognised that businesses are seeking to define their value proposition and business strategy based on the notion of product stewardship as shared value. By definition, the "adopt a stewardship role" archetype seeks to develop and support the network of stakeholders that sustains the value creation, and this can be achieved by maximising the positive environmental and social impacts associated with the value chain.

Nestlé is one example of a company that bases its environmental program, including product stewardship initiatives, on the idea of shared value creation. Their stated purpose is "enhancing quality of life and contributing to a healthier future. Our approach creates value not only for our business, but also for individuals and families, for our communities and for the planet."[14]

Participation in recycling programs is one source of shared value creation if it helps to secure access to raw materials or enhances corporate reputation. The former Director of Sustainability for Nestlé Waters North America supported EPR laws in the past because the company believed that it would increase the supply of recycled plastics for their beverage bottles.[15]

Updating Product Stewardship for a Circular Economy

As discussed above, product stewardship can support positive environmental and social benefits across a range of product categories from chemicals to batteries and packaging. Next, we consider the need to update product stewardship to meet the needs of a circular economy.

The circular economy concept represents a shift away from linear, disposable flows of materials, where raw materials are extracted, goods are produced, we use them, we collect and recycle some material, before disposal to landfill or incineration. In a circular economy, renewable flows of materials and energy are promoted through the deliberate redesign of products and systems. New value is created by cycling resources through the economy through multiple use-cycles to maximise the benefit to society through repair, reuse, remanufacturing, as well as recycling.

Circular economy thinking is inspired by earlier concepts and established academic disciplines such as industrial ecology. Industrial ecology emphasises the importance of renewable flows of resources and the concept of waste as a resource by drawing parallels between industrial and biological systems. Industrial ecology promotes a system-wide view of industrial processes and seeks to minimise waste and pollution and maximise value in the use of resources, energy, and capital inputs.[16]

While industrial ecology tends to focus on industrial systems and processes, the circular economy concept gives focus to products and services, and particularly the role of new business models in enabling new modes of consumption.[17] Examples of new business models that are promoted to achieve a circular economy are service models like renting, leasing, and sharing that move away from individual ownership models toward stewardship. In other words, when a business retains ownership of the product based on a leasing model, the business also retains the risks and responsibilities associated with the product throughout the product life cycle.[18] In this way, the circular economy concept aligns with the different meanings of product stewardship described above and product stewardship programs can play an important role in supporting the transition to a circular economy.

An emerging critique of the circular economy observes that, in the context of growing rates of consumption, despite the benefits of recycling materials to offset demand for primary resources, recycled materials cannot meet total demand. There is also often a lag between when materials that are in use become available as secondary materials to meet demand for new manufacturing,[19] and rapid technological advancement leads to demand for new material combinations that may limit the potential to manufacture new products using secondary materials.[20]

In the context of waste, a key sector where the circular economy concept is widely applied,[21] this means going beyond the conventional waste management approaches that aim to minimise disposal to landfill by prioritising the recovery of materials and energy. Whilst minimising disposal to landfill and maximising recovery is important, such an approach is not enough to bring about a sustainable circular economy where the positive impacts may be offset by the negative impacts of growing levels of consumption and production. There is also a risk of rebound effects where cost efficiencies gained from waste minimisation enable increases in outputs that undermine potential benefit.[22] Acknowledging this, it is apparent that the transition to a circular economy needs to go beyond the recovery of materials and energy and avoid excessive consumption and waste.

Limitations of Product Stewardship in a Circular Economy

While product stewardship can generate positive environmental outcomes as well as other benefits for those involved, it does not always deliver optimal outcomes for a circular economy. Despite early promise it is limited in its ability to:

Address increasing levels of consumption and waste;
Incentivize circular redesign;
Support reduction, repair, or reuse; and
Drive innovation at the ecosystem level.

Each of these issues is addressed below.

Addressing Increasing Levels of Consumption and Waste

A critical question is how product stewardship could support a circular economy that addresses broader societal challenges associated with increasing consumption and a perception that this is often excessive, i.e., to the extent that it cannot be justified by need.

Whether or not consumption is considered excessive in any particular context is a matter of judgement. There is no doubt, however, that production and consumption are growing at an unsustainable rate. According to a report from the Ellen Macarthur Foundation, production of plastics has increased twenty-fold since 1964 and is expected to almost quadruple by 2050.[23] Of the estimated 74 million tons of plastics consumed globally each year, around 40 percent is disposed to landfill and 32 percent leaks into the environment through illegal dumping, littering, or mismanagement.[24]

There is a growing body of evidence, for example, that plastics in the marine environment, particularly from single-use packaging and microfibers, are causing significant ecological damage[25] and EPR and other product stewardship schemes have been criticized for their lack of focus on marine pollution. Monroe argues that marine pollution must be addressed through improved information on product losses and integration into the goals and design of EPR schemes.[26]

While conventional EPR schemes have tended to focus on solid waste generated and disposed on land, some new initiatives at a corporate and industry level have started to address plastics in marine environments. While unlikely to have a major impact on reducing pollution, companies such as Adidas and Method have started to manufacture products from recycled marine plastics. The Alliance to End Plastic Waste, a global initiative funded by large corporations, funds projects that aim to improve waste management systems and reduce leakage of plastics into waterways and oceans.[27]

These initiatives are commendable, but they fail to address the problem at source, i.e., growing levels of consumption and waste. At present, there are limited examples of product stewardship schemes encouraging reduced consumption or absolute reductions in waste. Electrical and electronic products are a case in point. Despite the introduction of many industry-funded schemes to recover electrical and electronic waste ("e-waste") over the past two decades, global e-waste continues to grow and only 20 percent was recovered in 2016.[28] Factors driving increased consumption include a growing population, higher disposable incomes, growth in the digital economy, individuals owning multiple devices, and shorter replacement cycles for products like mobile phones and computers.

In grappling with how to justify and address growing levels of consumption and production in the context of a transition to the circular economy, Bocken and Short advocate for a "sufficiency-based circular economy."[29] They argue that sufficiency is emerging as a new paradigm that builds on the circular economy paradigm. While the circular economy promotes reuse and recycling aligned with a "making and doing more with less" approach, the sufficiency approach promotes a reduction in consumption and production aligned with "making do with less." As observed above, where the circular economy approach gives priority to the redesign of products and services and new business models, a broader and more fundamental society-wide change is necessary to transition to sufficiency. Moreover, a focus on sufficiency may be a necessary counter to the limits to a circular economy.

Incentivizing Circular Redesign

EPR and other product stewardship schemes have supported the development of waste management systems for a diverse range of waste streams, including packaging, e-waste and batteries. In practice, most of these schemes have been successful in supporting collection and recycling; however, they provide limited incentives for circular design and the shift to a circular economy by promoting reduction, repair, and reuse.[30]

A major study for the European Commission (EC) concluded that there was little evidence of a positive correlation between EPR policies and improvements in design.[31] The authors attributed this to a lack of targets or incentives for design and the fact that the fees paid by members of EPR schemes are generally based on average costs rather than costs associated with each individual producer's products. Some schemes *do* charge differential fees based on environmental impact or recyclability, however. In France, battery stewardship organisations are required to introduce "eco-modulated fees" for members based on a number of criteria, including the impacts of the battery chemistry, efficiency, the number of recharge cycles, risks, and relative impact at end of life.

More recently, waste policy in the European Union (EU) has shifted its focus toward circular economy principles with the adoption of the Circular Economy Action Plan in 2015[32] and the EC has been proactive in identifying ways to drive circular design through regulation. Best practice guidelines for EPR schemes identified some key principles, including:

> In line with the original goal of EPR, which is to foster eco-design and closed loop systems by having producers internalise the end-of-life management costs of their products, fees paid by each producer should reflect the actual end-of-life management costs of its own products as much as possible.[33]

The EC is considering options to update the rules on EPR schemes to reward the most sustainable design choices.[34] This is expected to provide guidance on eco-modulation of fees to provide a meaningful reward for sustainable product design choices.

Other policy initiatives are outlined in the EC's Circular Economy Action Plan in order to "make sustainable products, services and business models the norm and transform consumption patterns so that no waste is produced in the first place."[35] The Packaging Directive, for example, will be updated to include a stronger focus on design. Options that will be considered include targets for reducing "overpackaging" and waste, and restricting certain packaging materials where reusable alternatives are available or where consumer goods can be delivered without packaging.

Supporting Reduction, Repair, or Reuse

Circular economy models place a high priority on reuse and repair to extend the life of products before they need to be recycled, but these strategies are absent or disincentivized by many product stewardship schemes and regulations.

Currently, there is a lot of interest in the opportunity to reuse electric vehicle (EV) batteries, for example; however, the end-of-life management approach defined by the Battery Directive may inhibit broad uptake. The anticipated EV market worldwide corresponds with a dramatic increase in demand for lithium-ion batteries that will inevitably reach end of life. However, for EV applications end of life is defined when the battery's capacity has declined to 70–80 percent of the initial capacity, and, thus, the residual capacity can potentially be used in other applications before recycling. This alternative "end-of-first-life" option aligns with the circular economy concept because there is an opportunity to maintain the value of the product in the economy for longer, and waste generation is minimised. Already a range of stationary energy storage second-life applications are being explored, including for residential back-up power systems, or smart-grid applications whereby the battery provides power buffering to support the grid.[36] The reuse of used car batteries in forklift trucks and other factory vehicles is also being explored.[37]

Although second-life applications align with the goals of the Battery Directive, reuse remains undefined within the policy and there is a risk that the existing regulatory framework may limit second-life applications.[38] Redefining end of life to accommodate reuse within the policy must address how to share responsibility when ownership of the waste battery is transferred from the original producer to the supplier of the second-life battery. Equally, second-life performance needs to be better characterised, including expected duration of second life, to understand when the battery may be defined as a waste battery and available for recycling. Under the Circular Economy Action Plan, the European Commission has committed to

> revising EU consumer law to ensure that consumers receive trustworthy and relevant information on products at the point of sale, including

on their lifespan and on the availability of repair services, spare parts and repair manuals; and work towards establishing a new "right to repair" and consider rights for consumers with regards to availability of spare parts, or access to repair and upgrading services.[39]

These strategies strongly focus on electrical and electronic products. The EU Ecodesign Directive, for example, will be updated to require devices to be designed for durability, repairability, upgradability, maintenance, reuse, and recycling. There have also been attempts in the United States to introduce right to repair legislation – so far with limited success.[40]

Other examples focus on fashion, including initiatives led by individual companies. Patagonia, for example, promotes responsible consumption and repair through stores that provide a free repair and alteration service through what they call their Worn Wear® Repair Hub. The company argues that "One of the most responsible things we can do as a company is to make high-quality stuff that lasts for years and can be repaired, so you don't have to buy more of it."[41]

The examples provided so far include EPR policy development and company-led product stewardship initiatives; however, stewardship by consumers also needs to be acknowledged and supported, particularly in the context of repair and reuse. While consumers are encouraged to buy more sustainable products and packaging and to separate their wastes for recycling, they also play an active role in extending the life of products through formal and informal systems of reuse, exchange, and repair.[42] New business models that aim to "close the loop" through reuse or recycling place additional responsibilities on consumers: either to return products to the correct location; or in the case of leasing models, to ensure that products are returned in good condition. These models "require a level of care and stewardship of products well beyond traditional consumer–producer relationships."[43]

Driving Innovation at the Ecosystem Level

Extended responsibilities for consumers highlight the need to drive innovation beyond conventional value chains. EPR policies have tended to focus on the central role of the "producer" and their direct influence on product design and procurement, while references to product stewardship generally encompass a wider range of actors in the value chain. The Product Stewardship Institute in the U.S., for example, notes that "[t]he producer of the product has the greatest ability to minimise adverse impacts, but other stakeholders, such as suppliers, retailers and consumers, also play a role."[44]

There is some recognition, however, that transitioning to a circular economy requires innovation at the level of an "ecosystem" as well as product and business model innovation.⁴⁵ As an example, designing a product for reuse does not achieve circularity unless there are also fundamental structural changes enabling the deployment of a reusable product. Key changes will likely encompass new product design and manufacturing; a new business model, e.g., leasing, to capture the value in reusing; infrastructure to support reuse, including return logistics, cleaning, and repair; information and marketing campaigns to establish a customer base; new roles for product warranty to guarantee quality and reliability; and potential changes to consumer law. It is clear that a systems view is important.

EPR policies and product stewardship programs more broadly are well positioned to support ecosystem change to achieve circularity. Regulations can include sectors across the whole value chain and product stewardship programs can be designed to explicitly encourage broader participation beyond narrowly defined "producers." Product stewardship organisations (PSOs) can facilitate new partnerships and coordinate collective action; and stewardship programs can provide a safe space for experimentation supporting innovation where the risks, costs and impact may be shared.

The Australian Packaging Covenant Organisation is an unusual example of a PSO that operates explicitly within a "collective impact model." Their framework to achieve the 2025 national packaging targets states that

> Considering the complex packaging value chain, it is vital that stakeholders from different sectors commit to a common agenda to address the complex social, economic and environmental issue; stakeholders cannot work in isolation to solve these problems.⁴⁶

This approach is based on all participants having a shared vision for change, continuous communication across the many players to build trust, the "backbone" or coordinating function being delivered by APCO, mutually reinforcing activities being delivered through collaborative and strategically aligned projects, and a shared measurement system.⁴⁷

Conclusion

Many of the environmental challenges that we face can be linked to consumption, from the depletion of non-renewable raw materials to air and water pollution and climate change. Product stewardship policies and programs have been introduced to address the environmental impacts of products over their life cycle by placing more responsibility onto producers. EPR policies, in particular, were intended to drive changes in design to improve recyclability and to provide additional funding for recycling programs. While these programs have undoubtably helped to increase recycling for some targeted products, such as packaging, e-waste and batteries, consumption continues to increase and many products still have no end-of-life solution.

Product stewardship will continue to be important because it holds producers and others accountable for their impacts and helps to support recycling at end of life. The idea needs to be refreshed and updated, however, to address urgent challenges, such as climate change and marine pollution, which are closely linked to increasing levels of production and consumption.

Emerging circular economy discourses show a potential way forward and product stewardship policies could be important in supporting a transition to a circular economy that addresses these broader and urgent societal challenges. Product stewardship policies and individual schemes need to provide greater incentives for changes in design, including through eco-modulation of fees. They also need to broaden their scope to promote new business models involving reuse, remanufacturing, and shared consumption as well as recovery at end of life if we want to reduce our resource use and waste.

Another priority is to use product stewardship policies and programs as a vehicle to focus urgent attention on problems associated with plastics, and their impacts in the marine environment. "Wicked problems" such as these require more than shared responsibility along the value chain. They require a complete rethink of the way that product ecosystems are created and managed. Such radical changes can shift consumption behaviors and need to be supported by coordinated actions of government, industry, and citizens, and product stewardship policies and schemes could help to facilitate these change processes.

Notes

1 Helen Lewis, "Product Stewardship: Institutionalising Corporate Responsibility for Packaging in Australia" (PhD Diss., RMIT University, 2009).
2 Based on Lewis, "Product Stewardship," 25.
3 OECD, *Extended Producer Responsibility – Updated Guidance for Efficient Waste Management* (Paris: OECD Publishing, 2016), 16.
4 Lucas Porsch et al., *Development of Guidance on Extended Producer Responsibility* (Brussels: European Commission, 2014).
5 Daniel Kaffine and Patrick O'Reilly, *What Have We Learned about Extended Producer Responsibility in the Past Decade? A Survey of the Recent EPR literature* (Paris: OECD Publishing, 2013).
6 "Batteries and Accumulators," European Commission, accessed June 20, 2021, https://ec.europa.eu/environment/waste/batteries/.
7 "APCO's Co-Regulatory Model," Australian Packaging Covenant Organisation, accessed June 20, 2021, https://apco.org.au/apco-s-co-regulatory-model.
8 Lewis, "Product Stewardship," 25.
9 Apoorva Mandavilli, "The World's Worst Industrial Disaster is Still Unfolding," *The Atlantic*, July 10, 2018, https://www.theatlantic.com/science/

archive/2018/07/the-worlds-worst-industrial-disaster-is-still-unfolding/560726/.
10 Jean Bélanger et al., "Responsible Care: History and Development," in *Responsible Care: A Case Study*, eds. Jean Bélanger et al. (Berlin and Boston: De Gruyter, 2013).
11 Manfred Fleischer and Michael Troege, "Organising Product Stewardship in Large Chemical Companies." *Journal of Business Chemistry* 1, no. 2 (2004): 29.
12 Michael Porter and Mark Kramer, "Strategy and Society: the Link Between Competitive Advantage and Corporate Social Responsibility." *Harvard Business Review* 84, no. 12 (December 2006): 78–92.
13 Nancy Bocken, Samual William Short, P. Rana, and Steve Evans, "A Literature and Practice Review to Develop Sustainable Business Model Archetypes." *Journal of Cleaner Production* 65 (2014): 42–56.
14 "Our Commitments: Where We Have Impact," Nestlé, accessed April 9, 2020, https://www.nestle.com/csv/impact.
15 Helen Lewis, *Product Stewardship in Action: The Business Case for Life-Cycle Thinking* (Sheffield: Greenleaf Publishing, 2016): 24.
16 Thomas E. Graedel and Reid J. Lifset, "Industrial Ecology's First Decade," in Clift Roland and Angela Druckman, *Taking Stock of Industrial Ecology*, eds. (Cham: Springer, 2016).
17 "What is the Circular Economy?" Ellen MacArthur Foundation, accessed September 10, 2021, https://ellenmacarthurfoundation.org/topics/circular-economy-introduction/overview.
18 Walter R. Stahel, "Circular Economy: A New Relationship with our Goods and Materials Would Save Resources and Energy and Create Local Jobs." *Nature* 531 (2016).
19 Thomas E. Graedel, et al., *Recycling Rates of Metals-A Status Report* (Paris: UNEP, 2011).
20 Julian M. Allwood, "Squaring the Circular Economy: The Role of Recycling Within a Hierarchy of Material Management Strategies," in Ernst Worrell and Markus Reuter eds., *Handbook of recycling* (Waltham, MA: Elsevier, 2014).
21 See for example, "WRAP and the Circular Economy." WRAP, accessed June 20, 2021, http://www.wrap.org.uk/about-us/about/wrap-and-circular-economy; European Commission. *A New Circular Economy Action Plan for a Cleaner and More Competitive Europe* (Brussels: European Commission, 2020).
22 Trevor Zink and Roland Geyer, "Circular Economy Rebound." *Journal of Industrial Ecology* 21, no. 3 (2017): 593–602.
23 "New Plastics Economy: Rethinking the Future of Plastics," Ellen MacArthur Foundation, accessed September 10, 2021, https://ellenmacarthurfoundation.org/the-new-plastics-economy-rethinking-the-future-of-plastics-and-catalysing.
24 Ellen MacArthur Foundation, "New Plastics Economy: Rethinking the Future of Plastics," 26.
25 See for example, Mils Simon, Doris Knoblauch, and Maro Luisa Schulte, *No More Plastics in the Ocean* (Berlin: Adelphi, 2018); Lene Jensen, Cherie Motti, Anders Garm, Hemerson Tonin, and Frederieke Kroon, "Sources, Distribution and Fate of Microfibres on the Great Barrier Reef, Australia." *Scientific Reports* 9, (2019) Art. no: 9021.
26 Leila Monroe, "Tailoring Product Stewardship and Extended Producer Responsibility to Prevent Marine Pollution." *Tulane Environmental Law Journal* 27 (2014): 219–36.
27 "Our Work," Alliance to End Plastic Waste, accessed June 21, 2021. https://endplasticwaste.org/en/our-work.

28 Cornelis Peter Baldé, et al., *The Global E-Waste Monitor 2017* (Bonn/Geneva/Vienna: ITU, United Nations University, ISWA, 2017), 5.
29 Nancy Bocken and Samuel William Short, "Transforming Business Models: Towards a Sufficiency-based Circular Economy." *Environmental Innovation and Societal Transitions* 18 (2016): 41–61.
30 Leal Filho, et al., "An Overview of the Problems Posed by Plastic Products and the Role of Extended Producer Responsibility in Europe." *Journal of Cleaner Production* 214 (2019): 550–558; Victor Mitjanz Sanz et al., *Redesigning Producer Responsibility: A New EPR is Needed for a Circular Economy* (Amsterdam: Zero Waste Europe, 2015).
31 Porsch et al., *Development of Guidance on Extended Producer Responsibility*.
32 European Commission, *A New Circular Economy Action Plan for a Cleaner and More Competitive Europe* (Brussels: European Commission, 2020).
33 Porsch et al., *Development of Guidance on Extended Producer Responsibility*, 125.
34 European Commission, *A European Strategy for Plastics in a Circular Economy* (Brussels: European Commission, 2018).
35 European Commission, *A European Strategy for Plastics in a Circular Economy*, 3.
36 "Germany's Largest EV Battery-Powered Stationary Storage System Will Give Grid Flexibility." *Energy Storage News*, accessed June 21, 2021, https://www.energy-storage.news/news/storage-system-using-renault-ev-batteries-to-be-built-in-germany.
37 "Audi Installs Used Lithium-Ion Batteries in Factory Vehicles," *AudiMediaCenter*, accessed June 21, 2021, https://www.audi-mediacenter.com/en/press-releases/audi-installs-used-lithium-ion-batteries-in-factory-vehicles-11371.
38 Silvia Bobba et al., *Sustainability Assessment of Second Life Application of Automotive Batteries* (Luxembourg: Publications Office of the European Union, 2018).
39 European Commission, *A New Circular Economy Action Plan for a Cleaner and More Competitive Europe*, 5.
40 Joyce Costello, "Momentum for Repair in the EU and US," *E-Waste Watch*, July 30, 2019, https://ewastewatch.com.au/2019/07/30/right-to-repair-eu-us/.
41 "Better Than New," Patagonia, accessed June 20, 2020, https://www.patagonia.com.au/pages/worn-wear.
42 Ruth Lane and Matt Watson, "Stewardship of Things: The Radical Potential of Product Stewardship for Re-Framing Fesponsibilities and Felationships to Products and Materials." *Geoforum* 43 (2012).
43 Stefan Kaufman et al., *The Drivers, Barriers and Interventions that Influence Business Uptake of Circular Economy Approaches* (Melbourne: Monash University, 2020), 21.
44 "What is Product Stewardship?" Product Stewardship Institute, accessed May 31, 2020, https://www.productstewardship.us/page/Definitions.
45 Konietzko, Jan, Nancy Bocken, and Eric Hultink, "Circular System Innovation: An Initial Set of Principles." *Journal of Cleaner Production* 253 (2020).
46 Australian Packaging Covenant Organisation, *Our Packaging Future: a Collective Impact Framework to Achieve the 2025 National Packaging Targets* (Sydney: APCO, 2020), 11.
47 Australian Packaging Covenant Organisation. *Our Packaging Future*.

Bibliography

Alliance to End Plastic Waste. "Our Work." Accessed June 21, 2021. https://endplasticwaste.org/en/our-work.

Allwood, Julian M. "Squaring the Circular Economy: the Role of Recycling Within a Hierarchy of Material Management Strategies." In *Handbook of Recycling*, edited by Ernst Worrell and Markus Reuter, 445–477. Waltham, MA: Elsevier, 2014.

AudiMediaCenter. "Audi Installs Used Lithium-Ion Batteries in Factory Vehicles." Accessed June 21, 2021. https://www.audi-mediacenter.com/en/press-releases/audi-installs-used-lithium-ion-batteries-in-factory-vehicles-11371.

Australian Packaging Covenant Organisation. "APCO's Co-Regulatory Model." Accessed June 20, 2021. https://apco.org.au/apco-s-co-regulatory-model.

Australian Packaging Covenant Organisation. *Our Packaging Future: a Collective Impact Framework to Achieve the 2025 National Packaging Targets*. Sydney: APCO, 2020.

Baldé, Cornelis Peter, Vanessa Forti, Vanessa Gray, Ruediger Kuehr, and Paul Stegmann. *The Global E-Waste Monitor 2017*. Bonn/Geneva/Vienna: ITU, United Nations University, ISWA, 2017.

Bélanger, Jean, Peter Topalovic, Gail Krantzberg, and Joanne West. "Responsible Care: History and Development." In *Responsible Care: A Case Study*, edited by Jean Bélanger, Maria Topalovic, Joanne West, Bernard West, Peter Topalivic, 1–20. Berlin and Boston: De Gruyter, 2013.

Bobba, Silvia, Andreas Podias, Franco Di Persio, Maarten Messagie, Paolo Tecchio, Maria Anna Cusenza, Umberto Eynard, Fabrice Mathieux, and Adreas Pfrang. *Sustainability Assessment of Second Life Application of Automotive Batteries*. JRC Technical Report. Luxembourg: Publications Office of the European Union, 2018.

Bocken, Nancy and Samuel William Short. "Transforming Business Models: Towards a Sufficiency-Based Circular Economy." *Environmental Innovation and Societal Transitions* 18 (2016): 41–61.

Bocken, Nancy, Samuel William Short, P. Rana, and Steve Evans. "A Literature and Practice Review to Develop Sustainable Business Model Archetypes." *Journal of Cleaner Production* 65 (2014): 42–56.

Costello, Joyce. "Momentum for Repair in the EU and US." *E-Waste Watch*, July 30, 2019, https://ewastewatch.com.au/2019/07/30/right-to-repair-eu-us/.

Downes, Jenni. "The Planned National Waste Policy Won't Deliver a Truly Circular Economy." *The Conversation*, October 1, 2018. https://theconversation.com/the-planned-national-waste-policy-wont-deliver-a-truly-circular-economy-103908.

Ellen MacArthur Foundation. "What is the Circular Economy?" Accessed June 20, 2021. https://ellenmacarthurfoundation.org/topics/circular-economy-introduction/overview.

Ellen MacArthur Foundation. *New Plastics Economy: Rethinking the Future of Plastics*. 2016. https://ellenmacarthurfoundation.org/the-new-plastics-economy-rethinking-the-future-of-plastics-and-catalysing.

EnergyStorageNews. "Germany's Largest EV Battery-Powered Stationary Storage System Will Give Grid Flexibility." Accessed June 21, 2021. https://www.

energy-storage.news/news/storage-system-using-renault-ev-batteries-to-be-built-in-germany.
European Commission. "Batteries and Accumulators." Accessed June 20, 2021a. https://ec.europa.eu/environment/waste/batteries/.
European Commission. "Sustainability." Accessed June 20, 2021b. https://ec.europa.eu/growth/industry/sustainability/circular-economy_en.
European Commission. *A European Strategy for Plastics in a Circular Economy*. Brussels: European Commission, 2018.
European Commission. *A New Circular Economy Action Plan for a Cleaner and More Competitive Europe*. Brussels: European Commission, 2020.
Filho, Leal, Ulla Saari Walter, Mariia Fedoruk, Arvo Lital, Harri Moora, Marija Klöga, and Viktoria Voronova. "An Overview of the Problems Posed by Plastic Products and the Role of Extended Producer Responsibility in Europe." *Journal of Cleaner Production* 214 (2019): 550–558.
Fleischer, Manfred and Michael Troege. "Organising Product Stewardship in Large Chemical Companies." *Journal of Business Chemistry* 1, no. 2 (2004): 26–36.
Graedel, Thomas E., Julian Allwood, Jean-Pierre Birat, Barbara K. Reck, Scott F. Sibley, Guido Sonnemann, Matthias Buchert, and Christian Hagelüken. *Recycling Rates of Metals-A Status Report*. Paris: UNEP, 2011.
Graedel, Thomas E., and Reid J. Lifset. "Industrial Ecology's First Decade." In *Taking Stock of Industrial Ecology*, edited by Clift Roland and Angela Druckman, 3–20. Heidelberg; Cham: Springer, 2016.
International Council of Chemical Associations. "Product Stewardship Guidelines." Accessed June 20, 2021a. https://www.rcsk.sk/mix/Product_Stewardship_Guidelines_ICCA.pdf.
International Council of Chemical Associations. "Responsible Care." Accessed June 20, 2021b. https://www.icca-chem.org/responsible-care/.
Jensen, Lene, Cherie Motti, Anders Garm, Hemerson Tonin, and Frederieke Kroon. "Sources, Distribution and Fate of Microfibres on the Great Barrier Reef, Australia." *Scientific Reports* 9, (2019) Art. no: 9021.
Kaffine, Daniel, and Patrick O'Reilly. *What Have We Learned About Extended Producer Responsibility in the Past Decade? A Survey of the Recent EPR Literature*. Paris: OECD Publishing, 2013.
Kaufman, Stefan, Jim Curtis, Alexander Saeri, Breanne Kunstler, Peter Slattery, Abby Wild, and Peter Bragge. *The Drivers, Barriers and Interventions that Influence Business Uptake of Circular Economy Approaches*. Melbourne: Monash University, 2020.
Konietzko, Jan, Nancy Bocken, and Eric Hultink. "Circular System Innovation: An Initial Set of Principles." *Journal of Cleaner Production* 253 (2020).
Lane, Ruth, and Matt Watson. "Stewardship of Things: the Radical Potential of Product Stewardship for Re-Framing Responsibilities and Relationships to Products and Materials." *Geoforum* 43 (2012).
Lewis, Helen. "Product Stewardship: Institutionalising Corporate Responsibility for Packaging in Australia." PhD Diss., RMIT University, 2009.
Lewis, Helen. *Product Stewardship in Action: The Business Case for Life-Cycle Thinking*. Sheffield, UK: Greenleaf Publishing, 2016.

Mandavilli, Apoorva. 2018. "The World's Worst Industrial Disaster is Still Unfolding." *The Atlantic*, July 10, 2018, https://www.theatlantic.com/science/archive/2018/07/the-worlds-worst-industrial-disaster-is-still-unfolding/560726/.

Monroe, Leila. "Tailoring Product Stewardship and Extended Producer Responsibility to Prevent Marine Pollution." *Tulane Environmental Law Journal* 27 (2014): 219–236.

Nestlé. "Our Commitments: Where We Have Impact." Accessed April 9, 2020. https://www.nestle.com/csv/impact.

OECD. *Extended Producer Responsibility – Updated Guidance for Efficient Waste Management*. Paris: OECD Publishing, 2016.

Patagonia. "Better than New." Accessed June 20, 2020. https://www.patagonia.com.au/pages/worn-wear.

Porsch, Lucas, Stephen Sina, Martin Hirschnitz-Garbers, Max Grünig, Tanja Srebotnjal, Jenny Tröltzsch and Benjamin Boteler. *Development of Guidance on Extended Producer Responsibility*. Brussels: European Commission, 2014.

Porter, Michael, and Mark Kramer. "Strategy and Society: the Link Between Competitive Advantage and Corporate Social Responsibility." *Harvard Business Review* 84, no. 12 (December 2006): 78–92.

Product Stewardship Institute. "What is Product Stewardship?" Accessed May 31, 2020. https://www.productstewardship.us/page/Definitions.

Sanz, Victor Mitjanz, Elena Diez Rica, Eva Fernández Palacios, Adrià Medina Alsina, and Noelia Vázquez Mouriz. *Redesigning Producer Responsibility: A New EPR is Needed for a Circular Economy*. Amsterdam: Zero Waste Europe, 2015.

Simon, Mils, Doris Knoblauch, and Maro Luisa Schulte. *No More Plastics in the Ocean*. Berlin: Adelphi, 2018.

Stahel, Walter R. "Circular Economy: A New Relationship With Our Goods and Materials Would Save Resources and Energy and Create Local Jobs." *Nature* 531 (2016).

van Buren, Nicole, Marjolein Demmers, Rob van der Heijden, and Frank Witlox. "Towards a Circular Economy: The Role of Dutch Logistics Industries and Governments." *Sustainability* 8, no. 7 (2016).

WRAP. "WRAP and the Circular Economy." Accessed June 20, 2021. http://www.wrap.org.uk/about-us/about/wrap-and-circular-economy.

Zink, Trevor, and Roland Geyer. "Circular Economy Rebound." *Journal of Industrial Ecology* 21, no. 3 (2017): 593–602.

11 An Evolutionary Systems Theoretic Perspective on Global Stewardship

William M. Bowen

As the human population continues to grow, if the democratic, middle-class structures of the liberal state are to continue to improve the lived experience of increasing numbers of people, technology will have to advance, economies to grow, and the global political and economic system to further integrate. This will require effective governance and global stewardship of the present generation's inherited endowment of culture, knowledge, and resources. At the same time, certain challenges of the Anthropocene must be overcome, such as loss of biodiversity, global warming, persistent institutionalized inequality, and the risks of all-out nuclear war. The obstacles they present are, on the whole, unintended consequences of the characteristically large-scale, complex urban, technological, political, and economic systems human societies have constructed since our distant ancestors began to develop cultures in the Pleistocene.[1] Their basic causes lie in the combined factors of population increase, advancements in technology, and conventional (but wrong) ideas about human nature and the relationship between humans and the environment. The interactions among these factors over many thousands of years have combined to produce the Anthropocene, the current era in which aggregates of human behaviors, in combination with advanced technologies and errors in the thinking and attitudes of millions of people, have caused industrial societies to have an unavoidable and dominant influence on the global climate, environment, and other species.

Stewardship can be interpreted meaningfully in several different ways, each of which creates tensions and paradoxes.[2] But regardless of how it is interpreted, the success or failure of the responses of various societies in meeting the challenges of the Anthropocene will depend on aggregates of decisions and actions taken by multitudes of people and groups disbursed throughout the world, and their interactions with the decisions and actions of myriads of other people and groups. All of these decisions and actions will occur within the context of various individuals' interpretations of their own immediate decision situations.[3] None will possess anything remotely similar to omniscience about their decision situations, or themselves, much less how their actions might aggregate with those of others to affect the overall global-scale challenges. Their knowledge will be imperfect and they will seldom, if ever make fully informed and rational choices. They will act on the basis of their own perceptions of order in the world, and these perceptions will at least in part be constructed and ascribed to it by the active ordering structure of their own minds. In many ways the outcomes of their actions will be mutually interdependent with those of huge numbers of other people, some of whom they know and others not, and often they will not recognize this interdependence. Moreover, they will act in the absence of a widely shared language containing commonly understood and accepted terms with which to describe their own local situations or how they aggregate to the much larger-scale and unimaginably more complex, global-scale challenges of the Anthropocene. They will have wildly divergent ideas and views of the factors that must be considered to effectively and appropriately steward their inheritance.

Thus, it should come as no surprise that when it comes to making collective efforts to steward the present inheritance, the state of the art is dismal. Indeed, vast experience suggests that efforts to overcome large-scale, complex social and technical problems typically amount to little more than collective action taken on the basis of partially formulated images or representations of poorly understood systems. When rendered and entered into political processes, such representations tend to amount to collective actions that, regardless of intent, are at best partially successful and, at worst, bring disastrous unanticipated consequences. Examples from the twentieth century include Marxism-Leninism and National Socialism, both of which were put forth as solutions to what were then perceived to be the major social problems of the time, as if they were based upon definitive social postulates.

How is one to think and go about fostering effective governance and appropriate stewardship of the present generation's inherited endowment of culture, knowledge, and resources under such conditions? Some parts of the answer can be found in other essays in this volume. The motivating idea behind this one is that such thinking can benefit greatly first from an understanding of evolutionary systems theory; second, from an historically and scientifically sound understanding of the urban, technological, political, and economic systems through which the challenges of the era have been produced; and third, from a recognition that some things can be changed and some not.

The guiding ethical principle I propose for this approach is to act whenever possible to increase the number of choices available for present and future generations. This requires, among other things, protection and preservation of the ecosystems and natural resources upon which human societies depend, often through public policy designed to achieve such goals and priorities as harnessing resources efficiently, and promoting such values as moral autonomy, representative government, and respect for nature. The guiding systems' theoretic principle I propose is to govern societal systems using distributed, local, autonomous controls on individual and group behavior whenever possible, rather than centralized, hierarchical controls, as will be explained in the following sections. Effective and appropriate stewardship is not about reconstructing the past or forecasting the future, but rather about *creating the future through acting in the present in accordance with this ethical and this systems principle*. To thus act and respond successfully to the challenges posed by the Anthropocene requires not only willingness to change and adapt as needed, but also knowledge, understanding, and acceptance of those things in human societies that cannot be changed.

One of the core elements of acting in accordance with these principles involves avoiding unanticipated consequences. Systems thinking can help greatly in this regard. Unintended consequences tend to occur due to neglect on behalf of agents to fully recognize and adequately consider the full range of factors that interact within their systems to produce their decision situations. The generalized logic may be described as follows: If in any given system two factors in a decision situation are interdependent, they are correlated. So let us suppose aspects A and B within a given decision situation are correlated. Let us further suppose that, say, an agent wants to manipulate B so as to alter a future outcome C. B and C are known to have a causal relationship, and B can be manipulated and changed at the agent's will. This means that by altering the level of B, the agent can alter the level of C, thereby changing the outcome. But if s/he is not also cognizant of the correlation between A and B, in manipulating B to alter C one of the effects will be also, and unknowingly, to alter A. The change in A is indirect, and may not be recognized until much later, if ever. The change in A is an unintended consequence of manipulating B to alter C. If the consequences of changing A are significant in some other aspect of the system, and perhaps even seriously deleterious, the actions taken to manipulate B may unintentionally create other, new, future problems, some of which are as serious, if not more serious than the original one leading to the decision to manipulate B. Unintended consequences can become especially problematic when decisions made by agents with a great deal of social authority or influence have the power to affect the lifeways of numerous other individuals, groups, and/or members of the communities of other species.

For example, the demand for wind turbines has grown tremendously in recent years. The prospects for using wind turbines is widely believed to have the potential to help overcome some of the challenges of the Anthropocene by reducing the production of greenhouse gasses from burning fossil fuels to generate electricity. But what often gets overlooked is the fact that the cores of wind turbine blades are almost invariably made of balsa wood, due to its highly desirable and practically non-substitutable combination of strength and weight. Thus, one of the

An Evolutionary Systems Theoretic Perspective 215

unanticipated outcomes of the recently increased demand for wind turbines has been a corresponding increase in the demand for balsa wood. In turn, this increased demand for balsa wood has led to the unintended consequence of greatly accelerated depletion of balsa trees from the Amazon rainforest, and an associated exploitation and destruction of indigenous tribes by an international balsa wood mafia.[4]

One of the clear messages of systems thinking, as outlined in the following, is that human scope is limited. Human perceptions of the world should be treated not immaculately but as hypotheses. Moreover, we cannot rely upon even what scope we have except in situations in which we are intimately, sensitively, and mindfully engaged. Tyra Olstad aptly refers to this engagement as having a "sense of place."[5] Such thinking paradoxically brings both insight and the temptation to ignore it, and powerful vested interests in the status quo strongly encourage the latter. Yet when ignored or unduly discounted, problems occur and suffering ensues for humans and other species, much of which is unnecessary.

Evolutionary Systems Theory

Any scientifically sound explanation or understanding useful for guiding global stewardship requires at least three components. First, it must contain a theory about how the world is organized and how it works. Second, it must have methods for gathering information about the world with which to test the validity of existing theory and/or improve working hypotheses. Third, it requires a technology with which to translate the knowledge gained with theory and methods into action in order to achieve desired goals and priorities. These components can be based either in classical, reductionistic analysis or in evolutionary systems theory, two different schools of scientific tradition.[6]

The classical approach uses reductionistic analysis in which the phenomenon being investigated is first conceptually decomposed and "reduced" into its component parts for explanation and understanding. This very powerful, well-organized and developed approach has been the basis for accumulating scientific knowledge since the Renaissance. Evolutionary systems theory, on the other hand, bases explanation and understanding on the assumption that the phenomenon of interest is a unified, holistic "system" of mutually interdependent, interacting parts, none of which can be conceptually separated from the whole without fundamentally altering it. The systems are "evolutionary" when, in the face of a changing environment, they either survive, persist, and the relationships between their core factors and components remain stable through some period of time, or not. While evolutionary systems theory is still in its relatively early stages of development, it can nevertheless lead to explanations and understandings that are every bit as historically and scientifically sound as those produced by the classical approach. At the same time the systems theoretic approach is at times much more useful for understanding and avoiding unanticipated consequences within large-scale, complex systems such as those that produce the challenges of the Anthropocene.

In the classical approach, theory is based upon the assumption that all observable events and outcomes that occur in the world have identifiable natural causes. It assumes that the world is composed of many complicated structures and that every structure has many complicated parts that interact to cause observable events. Each structure can be taken apart and its individual pieces can be examined independently using the same "cause and effect" paradigm. Events are caused by the interactions among these many different structures. An investigator can discover how structures cause specific outcomes by identifying, observing, and gathering data about each structure's independent causal parts and using models to uncover the stable patterns behind the interactions. The technology, which is to say knowledge about cause-and-effect relationships, can be used to alter events that occur repetitively. Since events are caused by discoverable patterns that occur within identifiable structures, humans can build machines and instruments with which to alter the status of specific parts of structures that are known to cause specific events or outcome. Altering the status of these parts will thereby cause changes in the events or outcomes that occur. This approach has transformed almost the entire body of accumulated knowledge about the world and how the world works over the past few centuries.

On the other hand, the evolutionary systems theoretic approach stipulates that the observable features of the events and the phenomenon of interest have been influenced by multiple interdependent factors that compose the systems of interacting components or parts from which the events and phenomena emerge. But while the system is composed of its parts, it is not equal to the sum of the parts that compose it. Systems create outcomes that may be observed and used to define problems only when such systems are "in motion," i.e., when the parts that make up the system interact in time and space. The whole may be greater than the sum of the parts.

Systems are always defined by an observer, and thus depend to some extent upon the observer's perspective. They are products of the human mind, and so are mind-dependent. But their structures and processes must conform to observation and data to remain coherent. They often have hierarchical structures and are always contained within boundaries that differentiate between the system and its environment. The environment as such is composed of everything outside the boundaries of the system at the highest level. Each level of structure within the hierarchy can be described by identifying relations among events and factors at that level. Lower levels of the hierarchical structures are embedded in the higher levels in the sense that the nature of the function at each successively higher level becomes more broadly embracing than that at the lower level. All events and factors within each level connect with and influence each other. The patterns that emerge from their interactions can be observed at any given time, but the patterns are not necessarily linear and not necessarily stable through time.

The methods used in systems theory begin with stipulating the system boundaries, its components or factors, and the patterns in their interactions. All factors in the system that interact at a given level are identified, especially those with strong connections, along with the principal feedback loops that characterize how the factors influence each other at that level of the system. Each factor's degree of independent agency is assessed: in the large-scale, complex systems such as those that have produced the challenges of the Anthropocene, humans typically have wide agency whereas technical factors have much more limited agency. Once the system has been stipulated, data are gathered to describe the system's dynamics by observing it in motion. The level of analysis that corresponds to the outcome or problem of interest is then verified and changed, if necessary. The ways each agent exerts its agency are identified to guide how they influence the dynamics of the system at their level. A determination is made as to whether the overall pattern of behavior seen in interaction between agents is cooperative or competitive when the system is in motion. Influences between levels of the system are assessed in terms of how they alter the dominant patterns of interaction, and their outcomes, along with those from outside the system's environmental boundary.

Once a system's components have been identified and patterns of interaction have been explored, the findings can be used to develop controls with which to intervene in search of better outcomes. In systems theory, controls are implemented through the imposition of constraints on the factors imposed for the purpose of giving direction to the behavior of the system or its component parts. Among the controls that can be implemented, for example, are altering the levels of specific factors within the system, re-engineering how specific factors interact with each other, introducing or eliminating feedback loops, altering or creating sensors to send signals to perform actions by agents, introducing new influences from other levels of the system, and/or introducing new influences from outside of the system's environmental boundaries. While in comparison to the technologies available through the classical approach, the technologies of evolutionary systems theory are still relatively new and undeveloped, they are nevertheless scientifically sound in at least one important sense. That is, they are testable against data. They are stipulated by observers and treated as testable hypotheses to be refined as necessary to conform to the relevant data. They are based upon efforts to coherently explain the patterns observed in the data in theoretical terms, some of which are typically rooted in human behaviors.

On this basis, two conceptually distinct strategies are available for use in controlling the behavior of individuals and groups in large-scale, complex social systems. One is centralized, top-down control by hierarchical authority, and the other is distributed, bottom-up control by local, autonomous individuals and groups. The top-down approach starts with the initiative of an authority at the highest level of a hierarchy who makes decisions about what behaviors are and are not acceptable for individuals throughout the entire system. For example, when Xi Jinping, president of the Republic of China, decrees that that country will be carbon neutral by 2060, the entire social system has no choice but to respond. His authority is delegated and disseminated throughout lower levels in the hierarchy, to lesser authorities who are, to a greater or lesser extent, bound by his decisions. Noncompliant thoughts and behaviors at the lower levels meet with punishment, sometimes severe. This approach is relatively reliable, especially when for whatever reason the entire system as a single unit is faced with adapting to changes in its external environment.

The bottom-up approach, on the other hand, starts with voluntary changes in behavior by local individuals and groups who take their own initiatives and mutually adjust to one another and their circumstances organically as they interact at all levels of society. Key elements in bottom-up control strategies tend to include emphasis upon things such as the tremendous generative potential of cooperative, organic, self-reflexive human actions: private property rights and governments that impartially define and protect them; voluntary action and exchange between legally free individuals: trust, reciprocity, informal enforcement of norms, and improvements in flows of information through abundant and open communication channels. Insofar as it is possible to place control in the latter, it is likely more in keeping with the ethical and systems theoretic principles proposed earlier in this essay. Extensive experimental and field work with common pool resources provides compelling evidence for the potential for bottom-up responses to environmental and resource problems and issues.[7]

The use of systems thinking in considering the challenges of the Anthropocene reveals an almost paradoxical insight into the selection of control strategies. The insight stems from the fact that global warming, loss of biodiversity, persistent institutionalized inequality, the risks of all-out nuclear war, and the other challenges of the era threaten entire social systems at the highest levels, and even the entire planet. They therefore infer an imperative for system-wide changes of a sort seemingly most reliably provided by top-down, centralized command and control strategies. However, evolutionary systems theory indicates that bottom-up strategies significantly raise the chances of attaining the collective goals and priorities most conducive to overcoming these challenges.

There are several reasons for this. First, top-down strategies are almost invariably based upon the erroneous assumption that a single human mind or small group of minds of the authorities at the top of the hierarchy can conceive of all the interacting parts in a large-scale, complex system, and therefore have the cognitive competence necessary to fully comprehend the situation and make rational decisions for the whole of the system without unintended consequences.[8] But because this assumption is predictably erroneous, top-down control strategies tend to engender inefficiency, waste of resources, cost ineffectiveness, and other unforeseen and potentially catastrophic unanticipated consequences. A stark example from the twentieth century is China's Great Leap Forward (1958–1962), during which time approximately 35,000,000 people needlessly starved to death or were brutally slain under the ubiquitously unquestioned and mistaken assumption that the authorities at the highest levels of the hierarchy were cognitively competent to fully understand what was going on around the country.[9] Second, the dynamics of cooperation in risky collective decisions favor bottom-up decision-making and local agreements.[10] Third, top-down strategies tend to suppress individual freedoms and to impose potentially severe limits on the number of choices available for individuals and groups disbursed throughout lower levels of the hierarchy, thus inhibiting creativity and self-initiative. No social authority, for example, can successfully and reliably decree by executive fiat that innovations conducive to successful dissolution of anthropogenically-caused global warming problems shall be created.

Inherited Systems that Have Given Rise to the Challenges of the Anthropocene

At least three categories of systems have emerged from the evolution of human society since the Pleistocene. They have interacted and combined to bring the present challenges of the Anthropocene into being. All require historical and scientific knowledge and understanding with which to generate realistic and appropriate strategies for intervention in search of better outcomes. These include urban systems, technological systems, political and economic systems. While in the end these three categories are not totally separable, they can be analytically distinguished as a means of gaining specificity and clarity about what can be changed and what cannot, as well as for purposes of improving the interventions used to meet and overcome the challenges.

Urban Systems

Urban systems began with the watershed period in human natural history known as the Neolithic Revolution, which began at different times and places roughly around 10,000 years ago. During this period, hunter-gatherer societies started to become sedentary, domesticate animals, and cultivate crops. While the typical hunter-gatherer prior to this period had captured and used only about 2,000 kilocalories of energy per capita per day, the agricultural advancements made in this period increased energy consumption per capita to perhaps 12,000 kilocalories per day. On some estimates, this increased energy consumption started to increase the number of mouths possible to feed per km² from 0.05 people with the foraging system toward the 54 today, and, perhaps, 70 to 80 by 2050.[11] In turn, the newly developed capacity to feed more people enabled a tremendous overall increase in the size of the human population. It also made surplus food production possible, and this was a prerequisite for the development of cities, since city-dwellers cannot as a rule grow enough food to support themselves and their families. Humans thus took an early, big step that put us on a virtually irreversible path toward the development of ever-larger, and ever-more-complex urban systems. Irreversible, that is, without raising the specter of relatively sudden and drastic reductions in the size of the population.

The next watershed period was the Industrial Revolution, starting around 250 years ago in Great Britain. This period brought rising industrial through-put (increased inputs to and outputs from industrial production processes), increasing per capita production and consumption, growing mechanization and machine-based production in factories, mining, transport and agriculture, and increasing percentages of total populations living in urban centers. A key element again was better agricultural productivity. This improved productivity broke up a previously unbreakable and vicious cycle between increased fertility, population growth, more total food consumption, less available food per person, greater levels of starvation, and still further increases in numbers of babies to work the farm. This was also the period during which capitalism and science first emerged in Europe.

The Neolithic Revolution and the Industrial Revolution were springboards for a population explosion that greatly increased the size of the human population from an estimated 4,500,000 – 6,000,000 individuals throughout the planet on the eve of the Neolithic Revolution to around 7,900,000,000 today, and growing, an increase of over 1300 times in just 10,000 years.

As the human population grew, so did the number, scale, and complexity of the world's urban systems. About 400 generations ago, just prior to the Neolithic Revolution, the total human population on Earth was estimated to have been about 200,000 hunter-gatherer kin groups of approximately 24 individuals of various ages. By about 200 generations or 5000 years ago, the largest city in the world was evidently Uruk, Iraq, with a population of about 45,000.[12] About 150 generations ago, within about a century after Hammurabi assumed the Kingship of Babylon in 1792 BCE, Babylon became the largest city in the world with a population of approximately 65,000. In 100 CE, or about 77 generations ago, the largest city in the world was evidently Rome, with a population of somewhere around 1,000,000 people. By about 60 generations ago, in 500 CE, Rome had shrunk and Constantinople had become the largest city with a population of 450,000. London exceeded 5,000,000 only about five generations ago, sometime just before 1900 CE. Between one and two generations ago, which is to say sometime between 1965 and 1975, Tokyo became the first to have over 20,000,000 residents. And today, according to United Nations statistics, the planet has 30 megacities with populations of over 10,000,000. The United Nations Population Division projects that the human population will grow by another 3,000,000,000 people or so by the end of this century. Among the effects of this growth have been vastly increased demand for food, fresh water, energy, and natural resources. Today, population size increases also raise considerations of increases in the supply of residual waste such as carbon dioxide from industrial production.

The available archeological, historical, and ethnographic evidence suggests that since the Neolithic Revolution, human population growth has followed a discernible pattern that may be described as an "upward spiral."[13] The spiral began with the increased subsistence resources needed for additional members of the relatively small population. Hamlets grew into villages and villages into cities as long as the necessary food and other subsistence resources were available to meet the needs of the additional population, which is to say as long as the population was no larger than the carrying capacity of its environment. Populations grew as long as enough accessible water, food, and crucial resources for subsistence were available. But as population size neared the environmental carrying capacity, resources became depleted, and local populations had to either turn to less desirable and more costly alternatives, starve, migrate, trade, raid, or face increased pressure to innovate technologically to improve the productivity of those that were available. Faced with these alternatives, day-to-day problems maintaining local social systems predictably brought increased pressures to invent and deploy new technology, new ways to socially organize production, and/or new forms of cultural and political regulation. The effect of any of these was to expand the carrying capacity of the environment by making it possible to exploit previously inaccessible resources necessary for subsistence.

One of the major components of the upward spiral toward large-scale, complex urban systems has been a positive (mutually amplifying) feedback loop between population growth and technological innovation. That is, as population has grown, new technologies have been invented and deployed, and these have enabled the population to grow further, which has led to still new technologies. In this respect, technological innovation has expanded the carrying capacity of the environment. Then, with the expanded carrying capacity, it has been possible to support more people in the local environment and population growth has resumed again. Human populations have thus grown, technology advanced, more efficient exploitation of the resources required for subsistence become possible, populations grown further, technology advanced further, and still more efficient exploitation become possible. This is subsistence "intensification."[14] The domestication of plants and animals during the Neolithic Revolution constituted one such innovation that enabled subsistence intensification. The rise of industrial production of goods and services powered by fossil fuels during the Industrial Revolution constituted another. In each period, prior to the invention and widespread adoption of the new ways of doing things, and the corresponding extension of the carrying capacity, the population could grow only to the previously smaller carrying capacity, at which point starvation set in and the population stopped growing.

Urban systems have thus evolved and spread from countless thousands of individual hamlets, villages, and towns with small population organized into small local networks, to larger regional networks, and occasionally into yet larger networked regional systems supported by transportation and communication technologies. All along, whether in a small village or large city, the concentration of population in urban centers enabled specialized division of labor and enhanced levels of trade. When combined with the invention and deployment of new technology, rates of available resource productivity greatly increased and local economies grew, making it possible to meet the subsistence needs of additional people in ever-larger urban populations.

That is, up to a point. At the same time as populations grew and technology advanced, the evolving urban systems presented new challenges. The use of new technology to exploit available resources for larger populations almost invariably exacerbated the stress on the scarce, immediately available and relatively easy to acquire resources necessary for survival. When this stress occurred, the larger populations predictably turned to a greater variety of less and less efficient accessible resources, thus tending to deplete the buffer available for surviving in lean times. In turn, when the buffer got small enough, starvation became a palpable danger that demanded strategies to manage the risk. Today, for example, the question of where the food, water, and other resources necessary for the subsistence of the presently expected additional 3,000,000,000 people in the world remains unanswered and is of serious concern in agricultural research circles.[15] Even the current levels of population cannot be sustained without continued large-scale industrial and agricultural production and highly intensive industrial exploitation of natural resources.

Once agriculturally-based hunter-gatherer populations got large enough relative to the carrying capacity of their accessible environments, further population growth depended not only upon inventing and deploying new technologies to intensify agricultural production enough to meet the subsistence needs of the population, but also upon finding new ways to organize socially, divide labor, specialize, and trade. Urban systems emerged, organized, and integrated the roles of various individuals and groups, enabling subsistence intensification to increase still further beyond levels possible in any of the previously acephalous, hunter-gatherer family groups. Local urban economies grew, developed, and demanded new forms of social organization for dealing collectively with conflict and competition between families for prize resources, defending urban centers against raiding and warfare, conducting trade in regional networks and controlling the centralized stores of community food kept and managed for use in times of low agricultural production. Urban economies grew and expanded, and local populations increased as new forms of socially stratified organization emerged: some people became chiefs or kings; others became vicars, warriors, servants, tradespersons, or farmers. In each case, the growth and expansion instantiated a basic rule in the evolution of human societies: every expansion of the economy and resource base upon which a human society subsists, while solving subsistence problems for the population, also comes with corresponding opportunity for control, enabling leadership, and ultimately, self-aggrandizing elites.[16] As urban systems evolved and developed, social stratification emerged, and some people became rulers and others became ruled.

Technological Systems

The advancement of technology since the beginning of the Industrial Revolution has enabled enormous increases in the scale of urban and social systems. For the present worldwide network of urban systems to continue to function, hundreds of millions of people must have access to enough food, water, and shelter to live and meet their tissue needs each day. Millions must commute to work, so wave upon wave of vehicles must course through their streets. Thousands of tons of cargo must move in and out of their freight terminals. Millions of gallons of clean water must flow silently through their pipes while millions more are carried away as waste and treated or not. At the same time, vast amounts of power must be consumed to run cars, planes, homes, businesses and industries, lights, gadgets, data centers, farms, defense systems, and stuff for personal consumption. Millions of gigabytes of data must flow between computers. Like the essential subsystems and components that keep a human body running, each of these subsystems and components is vital to the functioning of the present global network of urban systems. It all rests on the deployment of technology – most of which has been invented since the Industrial Revolution; the consumption of energy and resources to power it all and keep it going; and the resulting economic growth and development.

Once a new technology is invented, it multiplies. After having produced tool B to work on objects in class A, people began to imagine ways to improve the tool and soon they manufacture tool C to use in making tools of class B. Next, tool D is invented and used to improve tools of class C, and so on. Thus, fueled by the division of labor, specialization, progress, the steam engine, the scientific method, mercantilism, and capitalism, innovative new scientific ideas and technologies drove the Industrial Revolution and brought hypertrophic growth in new knowledge, tools, techniques, machines, and technologies. One of the effects was to enable greater focus of attention and coordination of the sequences of behaviors in production, vastly increasing rates of technological innovation and productive output per unit of land, labor, capital, and energy input. As more innovations and technologies emerged, the multiplication of technology mushroomed the numbers of new products and technologies, and as a result more and newer technologies increasingly shaped what was produced, how labor and material resource inputs were combined, how subsistence was obtained and work was organized, where work was conducted, and who could perform the work.

Advances in technology have intensified demand for resources and increased the production of residual waste products by enabling larger and more integrated networks of urban systems and subsystems. Today, for instance, the raw cotton for a shirt grown in Alabama, in the United States, may be shipped to Malaysia to be prepared, spun, and woven into threads and fabrics; the fabrics shipped to Africa to be dyed and cut into sleeves, collars, pockets, and other shirt parts; the shirt parts shipped to China to be assembled into shirts; the shirts sent to Korea to be marketed then sent to the United States for retail consumption. Those shirts not consumed in the United States within a year or two might then be shipped down market to Brazil or Argentina to be sold for less. In this way, through advancements in technology, increased specialization of economic activity, and increased levels of trade and cooperation between the people at different locations around the globe, the scale of markets and the technological systems that support them have increased beyond comprehension. Technology has thus enabled not only the scale of urban systems to expand, but also the number of linkages between people and places to escalate far beyond the limited sphere of perception and limited cognitive capacity of any human being to fully fathom.

Yet, although the scale and complexity of the present global social and technical systems cannot be brought meaningfully within the purview of the human mind, technology continues to advance, intensification continues to expand carrying capacities, and human populations continue to grow. Hopefully, that is. Billions of people today already owe their continued sustenance to highly intensified agricultural production produced and delivered through large-scale global, social, and technical networks. And the additional billions of people expected in the near future will depend even further upon the as-yet unknown and uncertain new methods required for as-yet un-invented new technologies to enable further intensification of agriculture.[17] The necessary inventions, innovations, and technologies cannot be commanded into existence from above by a hierarchically positioned social authority. Rather, the potential to alleviate future food crises, especially among resource-poor families in rural areas around the globe, rests in the first instance squarely on the shoulders of the local researchers and other knowledge workers in labs and agricultural fields making efforts to realign food systems, and secondarily upon the ways governance decisions get made to control the large-scale, complex social and technical systems in which they are to be adopted.

Political and Economic Systems

Sometime after the Neolithic Revolution, as the spiral of population growth and technological progress continued upward, pressures to create new forms of social authority to collectively regulate and control behavior began. For example, regions containing relatively large populations with community storage or reciprocal feasting required leadership and social controls for collective organization and decision-making. Similarly, as populations grew larger and denser relative to their environment's carrying capacity, the level of competition for desirable land and resources tended to rise within the population, and levels of violence to increase. At times, it became necessary to construct and manage irrigation systems, organize and orchestrate large labor forces for such activities as building and forming cooperative alliances for purposes of trade, successful intergroup competition, and mutual defense. The maintenance of security and social stability required greater social control and integration. New forms of social organization thus emerged into political economic systems that provided ways to hierarchically control human behaviors and interactions. From the very beginning, these hierarchical controls posed problems that required compromise between localized, autonomous decision-making by free individual agents, such as those that had characterized the family groups and acephalous local groups of hunter-gatherers for tens of thousands of years, and the centrally based exercise of social authority by individuals on higher levels of the hierarchy. Individual people subject to these controls lost some of their own freedom and autonomy, but the reproductive and collective survival benefits for the overall society subject to the controls proved itself to outweigh these costs.

The earliest social arrangements in hunter-gatherer society had typically consisted of several relatively autonomous individuals and small families, and interactions between them were face-to-face in nature.[18] Then as subsistence intensification occurred with the advancement of technology, populations grew and reorganized in new political and economic systems of increasing scale and scope. Local kin groups gave rise to clans, chiefdoms, regional polities, city-states, and eventually nation-states, megacities and globalized bodies. Resource use intensified. Social stratification became necessary for management of central stores, creation and maintenance of regional networks, capital investment, and trade relationships.[19] Various forms of tribal organization gave way to the emergence of political offices, group leaders and big men, chiefdoms, kingdoms, and empires. Social institutions, such as the Code of Hammurabi, were created to codify the ever-changing systems of social control, and to provide written records with which to pass them from generation to generation. Eventually, in the twentieth century, such institutions included the somewhat self-regulating capitalist system of market exchange in the liberal state, such as is presently found in the United States and its Western allied nations, and the contending, centrally planned socialist system of top-down governmentally controlled society, such as is found in the Islamic states of Iran, Saudi Arabia, and Bangladesh, and, in an alternate form, the communist nation of China.

As technology continues to advance, and populations continue to grow, still newer forms of social organization and control are likely to emerge. Political and economic systems will undoubtedly continue to evolve and organize in new and unforeseen ways. They will undoubtedly continue to require compromise between localized, autonomous decision-making by free individual agents, and the exercise of centralized control by hierarchically positioned social authorities. As urban systems get larger and more complex, humanity is even today witnessing a widespread search for new governance structures capable of providing or enhancing security and competitive advantage in newly globalized, regional economic systems. In theory, institutions supportive of vigorous freely competitive markets, bureaucracies administered with a sense of reason and rationality, protection by governments of socially beneficial property rights and public security, a fair system of justice to protect some level of individual liberty, and widespread participation and engagement in public affairs should all combine to ensure optimum autonomy for the individual while also ensuring social order along with efficient and effective provision of public and private goods. But in reality, at times individuals and relatively small, local groups do not trust the fairness or legitimacy of the institutional order in any given political and economic system, and they are likely to band together in powerful special interest groups, and sometimes even in revolutionary organizations committed to the overthrow of the current institutional order. Extreme cases of this sort of disaffection may be observed in the "terrorist" organizations found in various parts of the world. For reasons of such disaffection, among others, political and economic institution-building and maintenance has become more and more important to the point at which today the well-being of future generations may well rest squarely upon it. Such well-being will be either accelerated or impeded by the enforcement of policies, laws, and regulations, as well as by the configuration, capacity, and flexibility of the institutions responsible for delineating the powers and functions available in the governance structures of networks of urban systems, and also for steering future urban, regional, national, and global-scale system evolution and development.

Global Stewardship Within Our Inherited Systems

The systems theoretic context for global stewardship is one in which billions of people all around the industrialized world today benefit from the evolved, highly complex, large-scale, highly integrated world order, and they have huge vested interests in making sure that the status quo continues as long as possible in its current state. Self-aggrandizing elites exploit the lower classes while the governments of the world fail to harness resources efficiently, or to promote such vital values as moral autonomy, self-initiative and self-responsibility, representative government, or respect for nature. Meanwhile, the subsistence needs of the world's human population gets met primarily through already-highly intensified agriculture that must be intensified still further to meet the subsistence needs of the additional human population expected to be born over the next few decades. As the population continues to grow, so does the scale of exploitation not only of people, but also of natural resources, energy consumption, and amounts of residuals and waste products from industrial production, such as greenhouse gasses. It is no wonder that the global nature of sustainability challenges continues to grow and the challenges to intensify.[20] From an evolutionary systems theoretic perspective, the seeds for this were planted by distant ancestors in the Pleistocene, and the seeds have sprouted and continued to grow and unfold even to the present. While the current generation's inherited systems enable the subsistence needs to be met for a large and growing segment of the population, they also perforce bring correlate downsides, including global climate change, persistent institutionalized inequality, reduction in global biodiversity, risks of nuclear devastation, and a host of other potentially catastrophic possibilities from large-scale systems failures.

Within this context, how is one to think and go about fostering appropriate stewardship of today's inherited endowment of culture, knowledge, and resources? Although systems thinking is by no means a panacea, it can definitely help. If nothing else, it provides a deeply logical, evidence-based foundation from which to develop an understanding of the urban, technological, political, and economic systems through which the challenges of the Anthropocene have emerged. But there is much more. This approach also supports the acquisition of insight into what can be changed and what cannot, and it informs efforts to avoid the unintended consequences of well-meaning but mistaken actions taken to change those things that cannot be changed. Basically, the use of systems theory holds the promise of enhancing the ability of those who know and understand it to recognize appropriate actions for preserving the assets inherited from the past while also creating the greatest possible number of choices for generations yet to come.

From an evolutionary systems perspective, the root causes of the challenges of the Anthropocene, and the associated demand for global stewardship, stem largely from the factors identified in the two postulates articulated by Thomas Malthus in his famous 1798 "Essay on the Principle of Population." These stipulate first that food (subsistence) is necessary to human existence, and second that the passion between the sexes is necessary and will remain nearly in its present state. Thus, these problems arguably stem from aggregates of entire generations' worth of individual and collective efforts to solve the problems of subsistence in growing populations, and perhaps even to experience abundance, together with the drive to reproduce as members of the human species. But Malthus' postulates certainly do not tell the entire story. The root causes also include the ability of human groups to socially interconnect, create knowledge and culture, and pass so much of their cultural inheritance and newly acquired knowledge on from generation to generation.[21] These factors have combined in such a way that masses of people over many generations have been able to construct urban, technological, political, and economic systems that are now unimaginably large-scale and complex. Johnson and Earle stated this succinctly:

> The emerging world of global economic integration, which liberal economists hope will encourage the democratic, middle class structures of the liberal state, is in reality a further development of the intensification, integration, and stratification that have always characterized social evolution.[22]

In the main, the large-scale, complex urban, technological, political, and economic systems that have been inherited by the present generation are necessary for the human population to sustain its current size, and perhaps to grow by another 3,000,000,000 people over the next few decades. The continued, collective survival of the world's current population of nearly 8,000,000,000 people would be impossible without the advancements in technological, political, and economic systems that began during the Industrial Revolution. The additional people will only add to the challenges. The terms commonly used to describe such conditions in systems theory are "lock-in" and "path dependence." For purposes at hand, they mean that people throughout industrial societies will continue to drive their cars; fly their airplanes; build, heat, and light their homes, farms, businesses, and industries; consume their gadgets; and produce the food and materials necessary to keep it all going. This will all require unimaginable amounts of energy and resources. People will continue on largely as they have in the past because their societies are endowed with an inherited culture and set of institutions that came out of the Agricultural Revolution, and the Industrial Revolution, and these define and bound their collective behaviors and strategies. The lessons of human natural history suggest that if industrial societies cease to produce the goods and services that enable all of this to be done, the resources necessary for the subsistence of the current population will deplete relative to demand, starvation will set in, and the human population will shrink, perhaps drastically. There is no going back without great cost.

The inherited urban, technological, political, and economic systems found throughout the world today are in many ways effectively autonomous. That is, they are as a whole far too large-scale and complex to be consciously and intentionally manipulated by individual human agents, or small groups, at least not without raising the specter of disastrous unintended consequences, some of which may be potentially as bad or even far worse than the problems the actions are intended to solve in the first place. No such systems can be comprehended in their entirety by any given single mind, or even small group of minds, much less intentionally manipulated to significantly alter their overall paths in preconceived and preplanned directions. Yet all such systems depend vitally upon their environment for the energy and resource inputs necessary for subsistence, as well as for the capacity to absorb the residuals necessary for continued existence. The tension this creates between the independence of the systems from the point of view of individual agents embedded within them, on one hand, and on the other, the dependence of the systems on their environments, raises the specter that individual agents and small groups embedded within the systems may become insensitive to their critical dependence on their environment.[23]

The existence of the large-scale, complex systems required to keep current levels of industrial production going thus entails the considerable risk that individuals and groups in societies develop insensitivity to those environmental conditions on which their lives, and the lives of countless future people around the globe critically depend. This is of particular concern insofar as the over-deployment of complexity as a means of sustaining a population has historically led to societal collapse.[24] One implication is that it may be rational for individuals and groups in social systems to seek to reduce the size and complexity of their systems as a means of sustaining themselves.[25] Another is that improvements in the behavior of individuals and groups in human social systems require improvements in the quality of human thought, knowledge, and knowledge utilization throughout the population, such as through better and more widespread understanding and application of systems theory, as well as the meanings, tensions, and ideals involved in global stewardship.

An evolutionary systems theoretic perspective thus implies at least two distinct sets of goals and priorities for global stewardship. The first has to do with the thought processes behind the knowledge used to inform and guide courses of action taken to respond to and cope with problems that arise throughout all levels of our large-scale, complex systems each day. These begin with recognizing and exposing some of the widespread errors in the conventional ideas, assumptions, thinking, and attitudes among large segments of populations. People who assume that the resources of the world are infinite, for example, or that if something is good for you then more of it is better, are unlikely to take the sorts of corrective actions necessary to effectively steward their inheritance. In the same way, some of the conventional yet wrongheaded ideas and assumptions about human nature and the relationship between humans and their systems and environments pose serious impediments to improvement. For example, the assumption that humans have dominion and can achieve mastery over nature easily lends itself to a disregard of the environmental factors upon which sustainable societies depend for their continued existence. Similarly, several of the commonly made "killer" assumptions about decision-making and the behavior of complex systems are conducive to little more than potentially cataclysmic system failure.[26] Dörner's prescriptions for recognizing and avoiding error in complex systems can go a long way toward helping to identify the roots of failure in the mistaken assumptions behind the small, seemingly perfectly sensible steps that unwittingly set the stage for disaster on down the road.[27]

The second implication has to do with the interaction of thoughts, ideas, knowledge, and actions, including especially their aggregation from individual-level manifestation to collective outcomes. Efforts to govern and redirect large-scale, complex systems are almost invariably implemented through collective human action; the ways this action is designed, organized, social authority constructed, and knowledge utilized are of the utmost importance. In part, the success with which interactions between individuals and societies improve will depend not only upon improvements in the production, preservation, transmission, and utilization of knowledge at the individual level, but also upon the prevailing characteristics of the cultures, organizations, and institutions through which individual-level knowledge is made manifest and aggregated into collective outcomes. This was emphasized by Nobel laureate Elinor Ostrom and her colleagues in their extensive field and experimental research on collective action in ecological systems. One of their general conclusions was that relatively small, local groups who have the authority to exercise control from below tend to be in the best position to act in ways that are consistent with the guiding principles of increasing the number of choices available for present and future generations. While this conclusion stands in almost paradoxical contrast with the increasingly large-scale networks that human societies have tended to develop throughout history and before, to recognize, understand, and act in ways that help to create and preserve diverse economic, political and social institutions is, in a systems theoretic perspective, a good place to start to improve resilience and sustainability.[28] One of the primary ways to start to overcome the challenges of the Anthropocene is thus to start with efforts to achieve better political consensus among public policy-makers and to foster more accurate and complete perceptions formed by educational institutions and diffused and acted upon by growing segments of populations around the globe.

Finally, this perspective points to a social dilemma in which recognition of increasingly large scale and complex human social systems strongly dissuades the very same intimate, sensitive, and mindful engagement most conducive to better outcomes. One perspective on the dilemma infers that it can complicate and exacerbate the difficulties in developing the sort of "sense of place" Tyra Olstad describes as being conducive to the construction of positive feedback loops and the realization of meaningful opportunity for stewardship.[29] Another perspective looks at it through the lens of systems theory, in which case it can appear to be a variety of the more general problem of how to foster mutualistic collaboration, cooperation and build "conditions where reciprocity, reputation, and trust can help to overcome the strong temptations of short-run self-interest."[30] Though, as with any social dilemma, there are no glib, simple, single solutions, among the keys to improvement are better direct person-to-person communication, recognition of the interdependence of individuals and groups (including the implications of collective failure), and the creation of cultural conventions, social norms, and institutions conducive to enhanced levels of cooperation.

Notes

1 William M. Bowen and Robert E. Gleeson, *The Evolution of Human Settlements: From Pleistocene Origins to Anthropocene Prospects* (Cham, Switzerland: Palgrave MacMillan, 2019).
2 Maria Tengö et al., "Stewardship in the Anthropocene: Meanings, Tensions, Futures," below 234–251.
3 Larry Kiser and Elinor Ostrom, "The Three Worlds of Action: A Metatheoretical Synthesis of Institutional Approaches," in *Pathologies of Urban Processes*, eds. Kingsley E. Haynes, Antoni Kuklinski and Olli Kultalahti (Finland: Finnpublishers, 1985), 73–105.
4 Anonymous, "The Wind-power Boom Set Off a Scramble for Balsa Wood in Ecuador," *The Economist* (January 30, 2021): 54.
5 Tyra Olstad, "Stewardship and Sense of Place: Assumptions and Ideals," above 13–28.
6 Robert E. Gleeson, "Notes on the Differences Between Traditional Analysis and Systems Thinking." Unpublished notes, UST293: Human Origins of Global Warming (Cleveland, OH: Cleveland State University, 2021).
7 Elinor Ostrom, *Understanding Institutional Diversity* (Princeton, NJ: Princeton University Press, 2005).
8 Fredric August von Hayek, *The Fatal Conceit: The Errors of Socialism* (London: Routledge, 1988).
9 Jisheng Yang, *Tombstone: The Great Chinese Famine 1958–1962* (New York: Farrar, Straus and Giroux, 2008).
10 Francisco C. Santos and Jorge M. Pacheco, "Risk of Collective Failure Provides and Escape From the Tragedy of the Commons," *Proceedings of the National Academy of Sciences* 108, no. 26 (2011): 10421–10425.
11 Jean-Pierre Bocquet-Appel, "When the World's Population Took Off: The Springboard of the Neolithic Demographic Transition," *Science* 333, no. 6042 (2011): 560.

12 The estimates cited are evidentiarily based on artifact data such as cemetery remains. See Colin McEvedy and Richard Jones, *Atlas of World Population History* and Morris, *Social Development* (Great Britain: Penguin Books and Allen Lane, 1978).
13 Allen W. Johnson and Timothy Earle, *The Evolution of Human Societies: From Foraging Group to Agrarian State,* 2nd edition (Stanford, California: Stanford University Press, 2000), 29.
14 Johnson and Earle, *The Evolution of Human Societies: From Foraging Group to Agrarian State,* 22–24.
15 Vernon W. Ruttan, "The Continuing Challenge of Food Production," *Environment* 42, no. 1 (2000): 25–30.
16 Johnson and Earle, *The Evolution of Human Societies: From Foraging Group to Agrarian State,* 29–32.
17 Ruttan, "The Continuing Challenge of Food Production," 25–30.
18 Johnson and Earle, *The Evolution of Human Societies: From Foraging Group to Agrarian State,* 41–45.
19 Johnson and Earle, *The Evolution of Human Societies,* 31.
20 Tengö et al, "Stewardship in the Anthropocene: Meanings, Tensions, Futures," below 234–251.
21 Joseph Henrich, *The Secret of Our Success: How Culture is Driving Human Evolution, Domesticating our Species, and Making Us Smarter* (Princeton, NJ: Princeton University Press, 2016).
22 Johnson and Earle, *The Evolution of Human Societies: From Foraging Group to Agrarian State,* 389.
23 Vladislav Valintinov, "The Complexity–Sustainability Trade-Off in Niklas Luhmann's Social Systems Theory," *Systems Research and Behavioral Science* 31 (2014): 14–22.
24 Joseph A. Tainter, "Problem Solving: Complexity, History, Sustainability," *Population and Environment: A Journal of Interdisciplinary Studies* 22 no. 1 (2000): 3–41.
25 Vladislav Valentinov and Lioudmila Chatalova, "Institutional Economics and Social Dilemmas: A Systems Theory Perspective," *Systems Research and Behavioral Science* 33 (2016): 138–149.
26 John N. Warfield, *Understanding Complexity: Thought and Behavior,* (Fairfax, VA: AJAR Publishing, 2002), 237–247.
27 Dörner, *The Logic of Failure: Recognizing and Avoiding Error in Complex Situations* (New York: Merloyd Lawrence, 1996), 185–199.
28 Elinor Ostrom, "Why Do We Need to Protect Institutional Diversity?" *European Political Science* 11 (2012): 128–147.
29 Olstad, "Stewardship and Sense of Place: Assumptions and Ideals," above 13–28.
30 Elinor Ostrom, "A Behavioral Approach to the Rational Choice Theory of Collective Action," *The American Political Science Review* 92, no. 1 (1997): 1–22.

Bibliography

Anon. "The Wind-power Boom Set Off a Scramble for Balsa Wood in Ecuador." *The Economist* (January 30, 2021): 54.
Bocquet-Appel, Jean-Pierre. "When the World's Population Took Off: The Springboard of the Neolithic Demographic Transition." *Science* 333, no. 6042 (2011): 560–561.

Bowen, William M. and Robert E. Gleeson. *The Evolution of Human Settlements: From Pleistocene Origins to Anthropocene Prospects*. Cham, Switzerland: Palgrave MacMillan, 2019.
Cook, Earl. "The Flow of Energy in an Industrial Society." *Scientific American* 225, no. 3 (1971): 134–147.
Dörner, Dietrich. *The Logic of Failure: Recognizing and Avoiding Error in Complex Situations*. New York: Merloyd Lawrence, 1996.
Hayek, Fredric August von. *The Fatal Conceit: The Errors of Socialism*. London: Routledge, 1988.
Henrich, Joseph. *The Secret of Our Success: How Culture is Driving Human Evolution, Domesticating our Species, and Making Us Smarter*. Princeton, NJ: Princeton University Press, 2016.
Johnson, Allen W. and Timothy Earle. *The Evolution of Human Societies: From Foraging Group to Agrarian State*. 2nd Edition. Stanford, CA: Stanford University Press, 2000.
Kiser, Larry and Elinor Ostrom. "The Three Worlds of Action: A Metatheoretical Synthesis of Institutional Approaches." In *Pathologies of Urban Processes*, edited by Kingsley E. Haynes, Antoni Kuklinski and Olli Kultalahti, 73–105. Finland: Finnpublishers, 1985.
McEvedy, Colin and Richard Jones. *Atlas of World Population History*. Harmondsworth: Penguin Books and Allen Lane, 1978.
Morris, Ian. *Social Development*. Palo Alto, CA: Stanford University, 2010. http://pzacad.pitzer.edu/~lyamane/ianmorris.pdf.
Ostrom, Elinor. "Why Do We Need to Protect Institutional Diversity?" *European Political Science* 11 (2012): 128–147.
Ostrom, Elinor. *Understanding Institutional Diversity*. Princeton, NJ: Princeton University Press, 2005.
Ostrom, Elinor. "A Behavioral Approach to the Rational Choice Theory of Collective Action." *The American Political Science Review* 92, no. 1 (1997): 1–22.
Ruttan, Vernon W. "The Continuing Challenge of Food Production." *Environment* 42, no. 1 (2000): 25–30.
Santos, Francisco C., and Jorge M. Pacheco. "Risk of Collective Failure Provides and Escape From the Tragedy of the Commons." *Proceedings of the National Academy of Sciences* 108, no. 26 (2011): 10421–10425.
Tainter, Joseph A. "Problem Solving: Complexity, History, Sustainability." *Population and Environment: A Journal of Interdisciplinary Studies* 22, no. 1 (2000): 3–41.
Valentinov, Vladislav and Lioudmila Chatalova. "Institutional Economics and Social Dilemmas: A Systems Theory Perspective." *Systems Research and Behavioral Science* 33 (2016): 138–149.
Valentinov, Vladislav. "The Complexity–Sustainability Trade-Off in Niklas Luhmann's Social Systems Theory." *Systems Research and Behavioral Science* 31 (2014): 14–22.
Warfield, John N. *Understanding Complexity: Thought and Behavior*. Fairfax, VA: AJAR Publishing, 2002.
Yang, Jisheng. *Tombstone: The Great Chinese Famine 1958–1962*. New York NY: Farrar, Straus and Giroux, 2008.

12 Stewardship in the Anthropocene
Meanings, Tensions, Futures

*Maria Tengö, Johan Enqvist, Simon West,
Uno Svedin, Vanessa A. Masterson, and
L. Jamila Haider*

We begin by inviting the reader to join us in a moment of time travel. The year is 1992, the place is Rio de Janeiro, and we are here to attend the "Earth Summit": The United Nations Conference on Environment and Development. It is a pivotal moment in the recognition of sustainability challenges, such as biodiversity loss and climate change, as distinctively global issues that require collective responses. In our narrative, we are drawing on personal experiences as well as the reflections by Bo Kjellén, the chief negotiator for Sweden.[1]

The organizers of the event have deliberately placed the formal interstate negotiations some 50 kilometers outside the city in a large conference center. Here all the negotiators and their staff assemble daily – sometimes until late in the evening. Meanwhile, we find the other "societal actors" with an interest in the negotiations in the heart of Rio, in a vast sea of tents stretching along the shoreline in Flamenco Park. We walk among the tents, described by the conference as a "once in a lifetime" image of the world in miniature, and we encounter NGOs, activist groups, universities and research networks, industrial and corporate representatives, labor unions, media organizations, and many others. We find a microcosm of parties interested and affected by the nascent sustainability agenda – reflecting innumerable groupings of individual, social, and organizational interests, from local to global. We are told that everyone is here in Rio because of a shared recognition of the interwoven relations between humanity and the environment, and a shared mission to refashion these relations in more sustainable ways. This collective vision is sometimes described as representing a stewardship approach.

However, the more we listen to the various actors present, the more we discern differences between them. There are differences with regard to their thematic focus. Some are more interested in climate challenges, others in toxic waste, and still others in biodiversity. There are also differences in institutional roles and connected priorities and ideas about how to proceed. This is demonstratively visible in the physical separation between, on the one side, the formal state delegations as represented in the UN structure and, on the other side, the strongly diverse and highly

DOI: 10.4324/9781003219064-17

engaged NGO groups of other actors. The formal negotiation arena is deliberately located by the organizers in a closed conference facility far away, disjunct and isolated from the vibrant NGO platform in the beautiful beach park in central Rio.

Among all these actors, some are arguing that it is vital to set up global institutional and legal frameworks – but others hold a skeptical view, evident in particular among the formal delegations. Still others – especially in the NGO sphere – are emphasizing the importance of nurturing grassroots civic and municipal initiatives (an approach that later became part of the Agenda 21 Plan of Action). There are also differences with regard to political commitments voiced by various actors. And finally, there are also different notions of responsibilities. Some argue that efforts toward sustainability should be equally distributed between "developed" and "developing" countries. Others suggest that the world is un-equal and even in-equitable – and thus the distribution of responsibilities for action should reflect this historical fact, in terms of expected efforts for the future. Furthermore, some advocate for more strongly engaging the corporate sector while others argue that the corporate sector's "growth fetish" is the root cause of unsustainability.[2] We who are present in Rio experience a deep ambiguity between, on the one hand, a broadly shared mission, and on the other, innumerable and to some extent irreconcilable differences between parties. This ambiguity is well captured by the Swedish chief negotiator Bo Kjellén at 4 a.m. on the final night of the negotiations by the words "it is all agreed – as far as possible."[3]

Fast-forward 30 years to the present day, and while recognition of the global nature of sustainability challenges has surely grown, the challenges themselves have only intensified.[4] In 2021, geophysical and social scientists will submit a formal proposal requesting the International Union of Geological Sciences to officially designate the "Anthropocene," a new geological epoch defined by human impacts on the earth system.[5] Advocates claim that the Anthropocene is a vital concept for drawing attention to the magnitude of the pressures that humans are currently exerting on the environment,[6] and for characterizing the unique globalized, industrialized, and hyper-connected circumstances of the twenty-first century.[7] This development is often linked to the image of a "Great Acceleration" with regard to economic activity and resource use since the mid-twentieth century, through which humanity (or, more precisely, a small fraction of it) has pushed the earth system outside the natural variability of the Holocene.[8] At the same time, however, some of the key axes of debate present in Rio in 1992 have re-emerged in widespread contestation about both the appropriateness of the Anthropocene concept[9] and its implications for the pursuit of sustainability.[10] In this context, the concept of stewardship has been reinvigorated within the emerging transdisciplinary field of sustainability science as a means of articulating desirable responses to Anthropocene sustainability challenges, for instance, in notions of planetary stewardship,[11] Earth stewardship,[12] and biosphere stewardship.[13] Such approaches enroll older connotations of stewardship alongside contemporary ideas of ecological "tipping points," planetary boundaries, and complex adaptive systems.[14]

These revised framings sit alongside multiple other uses of stewardship in a wide array of environment-related fields, and a huge variety of uses in policy and practice, including government legislation, corporate sustainability agendas, civic environmental initiatives, and social movements. This makes for a complicated research and action landscape where revised notions of stewardship provide new opportunities to address the challenges of the Anthropocene, while also bringing with them inherited and new paradoxes and tensions.

In this chapter, we unpack the different ways in which stewardship is currently being interpreted, mobilized, and used, and draw out the tensions and possibilities for novel responses to Anthropocene challenges. We begin by drawing on our previous work[15] to describe four prominent interpretations of stewardship in the academic literature from 1990 to 2016, and identify some broad, cross-cutting tensions that arise from these different meanings. We then exemplify how these tensions are currently playing out in practical uses and enactments of stewardship through three case studies: government-led landscape stewardship incentive schemes in South Africa[16]; volunteer-led urban stewardship in both Bengaluru, India[17] and New York[18]; and finally, one case covering corporate-led biosphere stewardship of the world's oceans.[19] We draw these disparate examples together by suggesting that stewardship can be understood as a "boundary object" – a conceptual tool that "enables collaboration and dialogue between different actors whilst allowing for differences in use and perception."[20] We then introduce care, knowledge, and agency as useful concepts to better navigate tensions, relate different uses, and nurture pluralistic and critically reflexive approaches to stewardship. We conclude with a discussion of the possibilities provided by the stewardship concept to generate more sustainable social-ecological relationships in the Anthropocene.

Stewardship Meanings and Tensions

An initial glance at the contemporary stewardship literature can result in confusion. For example, on the one hand, one may find stewardship characterized as "a set of normative values that private individuals may hold, and that entail perceived duties and obligations to carefully manage and use marine resources," [21] whilst on the other, one may read that if "[any] action creates net ecosystem service value above the baseline condition, it [will] be considered to embody environmental stewardship."[22] These contrasting definitions suggest quite radical differences in formulations of stewardship across different academic traditions and contexts of use.

Table 12.1 Four broad meanings of stewardship in the academic literature from 1990 to 2016

Stewardship Meaning	Description
Ethic	Stewardship is characterized as a set of moral or philosophical principles that describe obligations and responsibilities to take care of nature
Motivation	Stewardship is characterized in terms of personal or collective attitudes, traits or preferences that generate pro-environmental behavior
Action	Stewardship is characterized in terms of activities or interventions, including policy programs, management actions, and social movements, in pursuit of environmental or social-ecological benefits
Outcome	Stewardship is characterized in terms of the achievement of defined results, including increased populations of threatened species or improvements in human well-being

Adapted from Johan Enqvist et al., "Stewardship as a Boundary Object for Sustainability Research."

Nevertheless, it is possible to identify some general patterns in the stewardship literature as a whole.[23] In our own qualitative systematic literature review of the literature from 1990 to 2016, we identified four broad contemporary meanings of stewardship: *ethic, motivation, action,* and *outcome* (Table 12.1). We found that those adopting the *ethic* meaning use the stewardship concept to describe moral or philosophical principles, obligations, and responsibilities to take care of nature or, in more holistic approaches, social-ecological relationships.[24] By contrast, those employing the *motivation* meaning characterize stewardship more in terms of attitudes, traits, preferences, and behavioral predispositions that incline people to engage in "pro-environmental" behavior, such as litter-picking or eco-friendly shopping,[25] and tend to focus on more instrumental and strategic reasons for engagement, such as financial reward or a sense of social belonging. Those employing the *action* meaning treat stewardship in terms of the performance of specific activities, practices, or interventions in pursuit of environmental or social-ecological benefits, ranging from agri-environmental incentive schemes,[26] to scientific monitoring in national park management,[27] to participation in community activism.[28] And those using the *outcome* meaning tend to frame stewardship in terms of the achievement of defined results or outcomes – most often ecological (e.g., increases in species populations) but also social and economic outcomes such as poverty reduction.[29]

The coexistence of these multiple meanings in the stewardship literature reveals the contestation between perspectives that has always been a central part of the pursuit of sustainability. The four broad meanings reflect different starting points, thematic concerns, epistemological practices, and desired end-goals, and tensions therefore emerge within and among different uses of the stewardship concept. Three tensions are especially pertinent:

(i) **Source of individual commitment:** should the stewardship concept refer primarily to deeply felt ethical senses of obligation and responsibility (e.g., to describe indigenous philosophical perspectives emphasizing reciprocity between humans and nature), or can it also refer to the performance of certain behaviors that can be "nudged" through information campaigns or economic incentives?
(ii) **Nature of collective action:** should stewardship be used to describe the development of managerial "top-down" programs and policies such as corporate sustainability initiatives, or to describe civic-led, "bottom-up" social movements and change processes?
(iii) **Breadth of desired change:** should stewardship refer to the achievement of specific, narrowly defined goals (e.g., such as increases in the population of a threatened species), or to describe the transformation of entire social-ecological systems?

The multiple meanings of stewardship, and the tensions they generate, have produced many different kinds of stewardship in policy and practice. To get a better sense of the tensions and possibilities of the stewardship concept, we now explore how different uses of the stewardship concept are playing out in real-world situations.

Multiple Stewardships in Policy and Practice

In this section we explore the use of the stewardship concept in three different case studies: government-led approaches to landscape stewardship in South Africa; volunteer-led approaches to urban stewardship in Bengaluru (India) and New York (U.S.A.); and, finally, corporate-led biosphere stewardship of the world's oceans. The cases represent the use of stewardship to describe activities at different levels, with different thematic concerns, and across different time frames.

Government-Led Landscape Stewardship in South Africa

In South Africa, governmental conservation authorities have used the term "Biodiversity Stewardship" to describe an agreement with private landowners that designates the landowner as the custodian of biodiversity and ecosystem services on their land. This approach, representative of an "action-" or "outcome-"focused approach to stewardship, involves the government provision of fiscal incentives and technical assistance in return for voluntary commitments from landowners through various levels of participation to conserve biodiversity and sustainably use resources. Consequently, in South Africa there is often a narrow association of stewardship with a top-down managerial approach to systematic biodiversity conservation with expert-led designation of spatial priorities.[30]

The case highlights tensions around the source of individual commitment for stewardship: should the term be used to describe an ethical sense of responsibility, or to describe behavior that is incentivized externally? The Biodiversity Stewardship program functions by incentivizing compliance through fiscal incentives and technical support for conservation targets. The South African National Biodiversity Institute refers to this program as a "highly cost-effective mechanism for expanding protected areas," illustrating the neoliberal foundations of this approach to stewardship. Interestingly, tax incentives were viewed as just an added bonus for landowners who were initially motivated largely by a moral responsibility for nature, and long-term participation was not secured by financial incentives but by extension assistance and training.[31] Additionally, research also shows that practitioners are engaging with stewardship beyond the neoliberal incentive focus, with many expressing a view of stewardship as responsible use and care of nature, and as a balancing act between stewards' use of natural resources for agricultural production and their responsibility to protect and manage the wider ecosystem.[32]

The Biodiversity Stewardship case also reveals tensions around the breadth of desired change. The term stewardship is used very narrowly in South Africa[33] – to focus on contractual incentive schemes for biodiversity conservation outcomes on private land and to expand protected areas. The widespread adoption and relative success of South African National Biodiversity Institute's program have obscured other uses and formulations of the word stewardship.[34] 20 percent of practitioners in a study by Cockburn and colleagues[35] conflated the term "stewardship" with South African National Biodiversity Institute's Biodiversity Stewardship tool and incentives. This prosaic focus on conservation outcomes is particularly problematic due to the history of forced removals and exclusion of people from land set aside for conservation in colonial and Apartheid South Africa (and elsewhere) and therefore risks that stewardship becomes a new concept to disguise business as usual.[36] However, there is some emerging evidence of more holistic, integrated practices in this sector. This includes the work of practitioners to broaden the interpretation of stewardship in South Africa by integrating other tools,[37] incorporating broader goals of social inclusion[38] and a wider focus on "Earth Stewardship,"[39] including a process-focus to embrace the complexity of social-ecological systems.[40]

Civic-led Urban Stewardship in Bengaluru (India) and New York (U.S.A.)

A significant part of the stewardship literature focuses on volunteering programs and self-organized grassroots efforts to restore, conserve, manage, monitor, advocate for, and educate the public about the local environment.[41] Extensive efforts to map and assess such civic organizations in a series of North American cities have shown that groups as diverse as bird watchers, community gardeners, kayakers, youth educators, professional NGOs, and friendly neighbors describe themselves as stewards.[42] Their work typically extends beyond simply caring for nature to focus on influencing how other people interact and form relationships with their local living environments. This is particularly important in cities, where limited public spaces can restrict ecological literacy, but an intermingling of people of various backgrounds and interests can create opportunities for novel ways of identifying oneself as a steward.[43]

However, this also creates tensions in terms of the nature of collective action for stewardship. Cities are often rife with inequality and seeking to protect "nature" in rapidly changing urban landscapes can lead to green gentrification, where more influential residents are able to define what uses (and sometimes users) are allowed in public spaces. In Bengaluru, India, restorations of traditional water bodies have sometimes created exclusive access and even privatization. However, more recent citizen-led initiatives have been able to create more inclusive, multifunctional spaces where middle-class joggers and birdwatchers coexist with traditional cattle herders, fishermen, and worshippers.[44] Formal authorities have an important role to play here, as devolution of power to stewardship networks initiated from the bottom-up creates new responsibilities for top-down institutions. This includes monitoring to ensure democratic participation and equal access to the spaces and benefits that are created.

This leads us to another tension in urban environmental stewardship: between the embodied care for one's local community and environment, and the external incentives applied from city administrations to encourage residents to assist them in various (often unpaid) tasks. One oft-cited example is the initiative to plant one million trees in New York, U.S.A., where it quickly became apparent that trees needed maintenance and care to survive. Through educational activities and "stewardship programs," the project was able to harness the labor of thousands of residents to ensure that project goals were reached.[45] The term stewardship can both be empowering and identity-building, and act as a tool for co-opting and redirecting local engagement.

Corporate-Led Biosphere Stewardship of the World's Oceans

While civic engagement in local communities is a common use for the stewardship term, Earth or biosphere stewardship has increasingly been held up as a more globally relevant concept as a means of achieving a "safe operating space" for humanity.[46] For example, Folke and colleagues[47] show how a few "keystone" actors (in the form of transnational corporations) have a disproportionately large impact on earth system processes; they call for a "Corporate Biosphere Stewardship" where this influence is wielded for, rather than against, the biosphere. A specific example is the "SeaBOS" initiative, a science–business partnership where seafood companies are collaborating for ocean stewardship.[48] The project shows that there are strong motivations even for transnational corporations to act as Earth stewards for various reasons: care for the planet, self-interest (for example to sustain fish stocks into the future), and also to meet changing consumer demand which is increasingly guided by sustainability certification schemes (such as the Marine Stewardship Council) and affected by lobby groups and campaigns.

These different motivations exemplify the tension around the source of the commitment, since participation in SeaBOS is at least partially motivated from a sense of ethical responsibility but is in large part incentivized externally. Despite pledging stewardship action, the intrinsic motivation of transnational corporations is arguably still profit-driven. Schneider and colleagues point out that the ethical foundations of biosphere stewardship, as embodying care, lie in tension with the disproportionate concentration of market share and power that such companies hold, which is argued to perpetuate social inequality and also biosphere degradation.[49] Can these paradoxes be productively harnessed?

The case of corporate stewardship also captures the tension around the nature of collective action, where working with and within existing power holders may be an important determinant for how far stewardship efforts can reach in terms of altering strong drivers of environmental degradation. The SeaBOS initiative[50] represents an approach of a "top-down" stewardship style, in which external actors have mobilized key players in the seafood industry to act in ways that lead to more sustainable outcomes, with the aim to transform processes and therefore paving the way for a larger-scale breadth of change. This raises questions regarding the authenticity or virtue of being an "Earth steward" when compared to the civic-led stewardship described above. Is the prospect of achieving far-reaching change through transnational corporations' vast operations more important for stewardship, than basing the initiative on more "genuine," bottom-up change processes? Or can they be complementary?

Stewardship as a Boundary Object

These examples from around the world provide a snapshot of some current ways the stewardship concept is being used. One response to this diversity might be to suggest that the concept should be more narrowly specified and used more sparingly. However, the existence of multiple kinds of stewardships can also be seen as evidence of the concept's broad and continuing appeal. We suggest that it is useful to think of stewardship as a "boundary object" for discussing pathways for sustainability in a complex world.

The term "boundary object" originates from Science and Technology Studies and describes a concept, framework, or tool that is "both plastic enough to adapt to local needs and constraints of the several parties employing them, yet robust enough to maintain a common identity."[51] The plasticity of boundary objects enables them to facilitate communication between many different parties around a shared topic or concern.[52] While a concise consensus definition may not be desirable for a boundary object, it can be useful to map out their contours, in order to better connect and make sense of different uses and interpretations. We present a framework for better understanding stewardship as a boundary object based on three connecting dimensions – care, knowledge, and agency (Figure 12.1).

We use the term "care" to connect the "ethic" and "motivation" meanings of stewardship. Care refers to the impulse or desire to "look after"

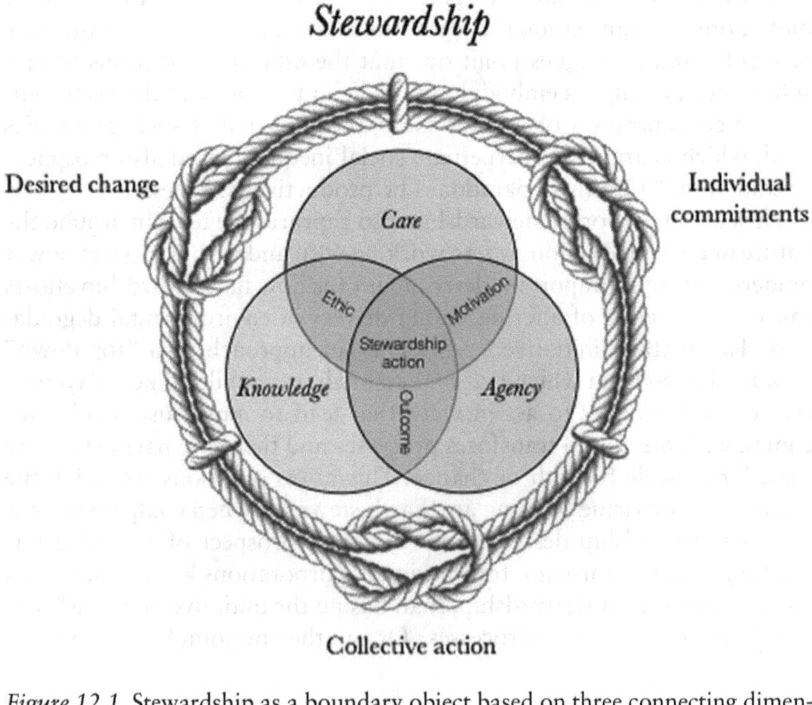

Figure 12.1 Stewardship as a boundary object based on three connecting dimensions – care, knowledge and agency (adapted from Enqvist et al. 2018). The three dimensions connect the different uses of the term stewardships as found in different sets of literature: as ethic, motivation, outcome, or as action. Outside the circle we indicate three tensions around application of stewardship that is addressed in this chapter.

something, on the basis of an "attentive interest and concern for its well-being."[53] Care helps us to better address tensions around the "sources of individual commitment" in cases of stewardship. A focus on care enables us to step away from binaries of *either* a deeply felt ethical responsibility *or* a simple financial or political motivation, and instead recognize the complexities of engagement around "stewardship" in particular situations, for particular purposes. For instance, in the South African case, many farmers viewed the economic incentives from government as merely a "bonus" for activities that they were already ethically committed to, so ethics and motivations seem to co-exist in the performance of stewardship. In the New York case, citizens were inspired to care for trees planted as part of a city government program, collaborating with authorities not because of economic incentives but because the work coincided with their existing ethical commitments. The global oceans case represents a tricky situation where researchers are actively attempting to support a "culture shift" whereby transnational corporations accept more ethical responsibility for their destructive behavior toward the biosphere, while at the same time recognizing that the performance of such a culture shift on the part of the companies (whether genuine or not) also potentially represents a competitive advantage in the market. The notion of care helps to recognize the distributed and collective nature of these different types of concerted stewardship action, recognizing the convergences of various ethics, motivations, and incentives among different parties, and the evolving nature of these different aspects through time.

We use the term "knowledge" to connect the "ethic" and "outcome" meanings of stewardship. Knowledge refers to the "information, know-how and ways of knowing that inform different types of stewardship action."[54] Tensions between ethic and outcome of stewardship are visible in the breadth of desired change, placing different demands on stewardship– e.g., holistic shifts in thinking across whole systems, or the delivery of measurable environmental benefits. Knowledge helps navigate this by showing how different demands are associated with distinct "ways of knowing" complex social-ecological change; understanding these cultural and epistemological associations can guide decision-making. The humanities and social sciences tend to emphasize deliberation around the nature of and reasons for stewardship as an ethic, while the natural sciences often focus on measurement of defined stewardship outcomes. A focus on complementary knowledge and processes for joint learning and sharing, translating, and reconciling different ways of knowing and acting can support mutual understanding and joint platforms for moving toward more productive enactments of stewardship.[55]

For example, in the South African case, early ideas of stewardship, informed by scientists working with biodiversity organizations, are increasingly becoming infused with greater insights from the social and complexity sciences – not least to broaden the appeal, resonance, and significance of stewardship among a broader range of social groups. In the global oceans case, there have been vibrant discussions between those researchers who see collaboration with transnational corporations as a pragmatic way of ensuring measurable outcomes for the biosphere, and those who see such collaboration as perpetuating deeper forms of social-ecological harm.[56] A focus on "knowledge" helps to recognize the value of different ways of knowing for improving such deliberation about the appropriate "means and ends" of stewardship.

We use the term "agency" to link the "motivation" and "outcome" meanings of stewardship. Agency refers to "the abilities and capacities of individuals, groups and organizations to engage in (collective) action and affect change."[57] Agency helps us to better address the interests and motivations for different kinds of actors, and the tensions around the "nature of collective action" – again moving away from binaries of *either* top-down *or* bottom-up action, and toward recognition of the complex convergences of agency that characterize enactments of stewardship. For example, the Bengaluru case shows how different constellations of actors have engaged in stewardship efforts that have had different impact on the abilities of disadvantaged groups to engage in action. In the global oceans case, researchers have attempted to engage with transnational corporations because of their apparent power to affect change, but at the same time the actual collaborative work has occurred with individual representatives who may have varying capacity to leverage systemic organizational transformation. The notion of agency helps to reveal the dynamic interplay between actors and forms of agency that converge within forms of stewardship to produce certain kinds of change.

Concluding Discussion: Stewardship and the Challenges of the Anthropocene

The key issue that we have grappled with in this chapter is how to handle the plural cognitive and philosophical bases of the stewardship concept – and at the same time be mindful of the continuously evolving meanings in the real-world contexts in which it is applied. Among all such meanings there is a distinct normativity to the word "stewardship" as a means of outlining and characterizing desired sustainability developments. This normativity is also embedded in existing institutions, processes, and platforms, setting the preconditions for the design of new and creative solutions and ways forward for sustainability – and in connecting biophysical, social, and cultural aspects of reality. We have explored these issues through examples and reflections around tensions and inherent paradoxes in plural usage of stewardship.

Stewardship – as well as its cognitive backgrounds – is an important contribution to the bioethical discussions of today. It is operationalized in a wide distribution of weak to strong positions about what it means for societal action to address complex social-ecological challenges. This could also be discussed in terms of ethical considerations for sustainability practice in terms of explicit or implicit norms.[58] An associated concern is about how normative perspectives relate to issues of complexity and the interwovenness of humans-and-nature. Recent work drawing on relational thinking and philosophy allows for a more thorough understanding of social-ecological phenomena as one inseparable totality and what this entails for handling sustainability challenges.[59]

The multiple meanings of stewardship are particularly salient as we face the emerging challenges of the Anthropocene. As our human presence has reached such a substantial level of environmental impact, then so too must our collective responsibility for mitigating these changes. Formulating the normative basis and institutional conditions for such responses calls for joint reflections on our collective and plural approaches and evidence bases of stewardship across the globe. It is among these connections between diverse human experiences – embodied within our accumulated knowledge and experience globally – that we might find sources of inspiration in navigating the Anthropocene era. To do so, it is critical to not only identify what may come harmoniously together, but also to be aware and mindful of the tensions and paradoxes emerging from the plurality of assembled experiences. In this context we have offered our framework for better understanding stewardship as a boundary object based on three connecting dimensions – care, knowledge, and agency. We hope that this framework, and other similar approaches, may stimulate thoughtful communication to draw out and examine the tensions and possibilities for novel responses to the unique and interconnected challenges of the Anthropocene. We have highlighted some avenues for finding appropriate paths for humanity at this stage of history and at the same time matching the quickly changing circumstances as they emerge in the pressing sustainability challenges of our times.

Notes

1 Bo Kjellén, "Diplomacy and Governance for Sustainability in a Partially Globalised World" (London: Routledge, 2014); Bo Kjellén, *A New Diplomacy for Sustainable Development* (London: Routledge, 2004).
2 Clive Hamilton, *Growth Fetish* (Sydney: Allen & Unwin, 2003).
3 Bo Kjellén, personal communication.
4 Will Steffen et al., "Planetary Boundaries," *Science* 347, no. 6223 (2015).
5 Meera Subramanian, "Anthropocene Now," *Nature News*, accessed 9 September 2021, https://www.nature.com/articles/d41586-019-01641-5.
6 Will Steffen, et al., "The Anthropocene: Are Humans Now Overwhelming the Great Forces of Nature?" *Ambio* 36, no. 8 (2007).
7 Victor Galaz, *Global Environmental Governance, Technology and Politics: The Anthropocene Gap* (Cheltenham: Edward Elgar, 2014).
8 Will Steffen et al., "Planetary Boundaries."

9 Yadvinder Malhi, "The Concept of the Anthropocene," *Annual Review of Environment and Resources* 42 (2017).
10 Peter Kareiva, et al. "Conservation in the Anthropocene," *Breakthrough Journal*, February 2012; Michael Soulé, "The 'New Conservation," in *Keeping the Wild* (Washington, DC: Island Press, 2014); Carl Folke et al., "Transnational Corporations and the Challenge of Biosphere Stewardship," *Nature Ecology & Evolution* 3, no. 10 (2019); Amselm Schneider et al., "Can Transnational Corporations Leverage Systemic Change towards a 'Sustainable' Future?" *Nature Ecology and Evolution* 4, no. 4 (2020).
11 Will Steffen et al., "The Anthropocene: From Global Change to Planetary Stewardship," *Ambio* 40 (2011).
12 F. Stewart Chapin et al., "Earth Stewardship," *Ecosphere* 2, no. 8 (2011).
13 Carl Folke et al., "Reconnecting to the Biosphere," *Ambio* 40, no. 7 (2011).
14 Carl Folke et al., "Social-Ecological Resilience and Biosphere-Based Sustainability Science," *Ecology and Society* 21, no. 3 (2016).
15 Johan P. Enqvist et al., "Stewardship as a Boundary Object for Sustainability Research," *Landscape and Urban Planning* 179, no. 2 (2018).
16 Jessica Cockburn et al., "The Meaning and Practice of Stewardship in South Africa," *South African Journal of Science* 115, nos 5–6 (2019).
17 Johan P. Enqvist, et al, "Against the Current," *Sustainability Science* 11, no. 6 (2016).
18 Heather McMillen et al., "Biocultural Stewardship, Indigenous and Local Ecological Knowledge, and the Urban Crucible," *Ecology & Society* 25, no. 2 (2020).
19 Henrik Österblom et al., "Emergence of a Global Science–Business Initiative for Ocean Stewardship," *Proceedings of the National Academy of Sciences of the United States of America* 114, no. 34 (August 22, 2017).
20 Enqvist et al., "Stewardship as a Boundary Object for Sustainability Research," 17.
21 Ingrid van Putten et al., "Individual Transferable Quota Contribution to Environmental Stewardship," *Ecology & Society* 19, no. 2 (2014): 1.
22 Joseph Nicolette, et al., "A Practical Approach for Demonstrating Environmental Sustainability and Stewardship through a Net Ecosystem Service Analysis." *Sustainability* 5 (2013): 2152.
23 E.g. Raphaël Mathevet, et al., "The Concept of Stewardship in Sustainability Science and Conservation Biology," *Biological Conservation* 217 (2018); Nathan Bennett et al., "Environmental Stewardship," *Environmental Management* 61, no. 4 (2018).
24 E.g. John Seamer, "Human Stewardship and Animal Welfare," *Applied Animal Science* 59 (1998); Jennifer Welchman, "A Defence of Environmental Stewardship," *Environmental Values* 21 (2012).
25 E.g. Angela Gupta, et al., "4-H and Forestry Afterschool Clubs," *Journal of Extension* 50, no. 3 (2012); Elizabeth Lokocz, et al., "Landscape and Urban Planning Motivations for Land Protection and Stewardship," *Landscape and Urban Planning* 99 (2011).
26 Sam Amy et al., "Hedgerow Rejuvenation Management," *Basic and Applied Ecology* 16 (2015).
27 Gary Davis, "National Park Stewardship and 'Vital Signs' Monitoring," *Aquatic Conservation: Marine and Freshwater Ecosystems* 15 (2005).
28 van Riper, "Women as Collaborative Leaders on Rangelands in the Western United States."
29 Dora Vega and Rodney Keenan, "Land Use Policy Agents or Stewards in Community Forestry Enterprises?" *Land Use Policy* 52 (2016).

30 Jaco Barendse et al., "A Broader View of Stewardship to Achieve Conservation and Sustainability Goals in South Africa," *South African Journal of Science* 112, nos. 5–6 (2016).
31 Matthew Selinske et al., "Locating Financial Incentives among Diverse Motivations for Long-Term Private Land Conservation," *Ecology and Society* 22, no. 2 (2017).
32 Cockburn et al., "The Meaning and Practice of Stewardship in South Africa."
33 Barendse et al., "A Broader View of Stewardship."
34 Barendse et al., "A Broader View of Stewardship."
35 Cockburn et al., "The Meaning and Practice of Stewardship in South Africa."
36 Cockburn et al., "The Meaning and Practice of Stewardship in South Africa."
37 Cockburn et al., "The Meaning and Practice of Stewardship in South Africa."
38 Yashwant Rawat, "Sustainable Biodiversity Stewardship and Inclusive Development in South Africa." *Current Opinion in Environmental Sustainability* 24 (2017).
39 Barendse et al., "A Broader View of Stewardship."
40 Cockburn et al., "A Relational Approach to Landscape Stewardship,"; Jessica Cockburn et al., "The Meaning and Practice of Stewardship in South Africa," *Land* 9, 224 (2020).
41 Erika Svendsen and Lindsay Campbell, "Urban Ecological Stewardship," *Cities and the Environment* 1, no. 1 (2008); Stanley Asah and Dale Blahna, "Motivational Functionalism and Urban Conservation Stewardship," *Conservation Letters* 5, no. 6 (2012); James Connolly et al., "Organizing Urban Ecosystem Services," *Landscape and Urban Planning* 109, no. 1 (2013).
42 Erika Svendsen et al., "Stewardship Mapping and Assessment Project," accessed 9 September 2021, https://www.nrs.fs.fed.us/pubs/50447; Johan Enqvist et al., "Place Meanings on the Urban Waterfront," *Sustainability Science* 14, no. 3 (2019).
43 Erik Andersson et al., "Reconnecting Cities to the Biosphere," *Ambio* 43, no. 4 (May 2014); McMillen et al., "Biocultural Stewardship."
44 Hita Unnikrishnan and Harini Nagendra, "Privatizing the Commons: Impact on Ecosystem Services in Bangalore's Lakes" *Urban Ecosystems* 18 (2015); Enqvist, et al., "Against the Current."
45 Campbell, "Constructing New York City's Urban Forest," in *Urban Forests, Trees and Greenspace. A Policy Perspective*, Anders L. Sandberg, Adrina Bardekjian, and Sadia Butt eds. (New York City, NY: Routledge, 2014).
46 F. Stewart Chapin et al., "Ecosystem Stewardship," *Ecosphere* 2, no. 8 (2011); Steffen et al., "Planetary Boundaries."
47 Folke et al., "Transnational Corporations."
48 Folke et al., "Transnational Corporations."
49 Schneider et al., "Can Transnational Corporations Leverage Systemic Change towards a 'Sustainable' Future?"
50 Österblom et al., "Emergence of a Global Science–Business Initiative for Ocean Stewardship."
51 Susan Leigh Star and James L. Griesemer, "Institutional Ecology, 'Translations' and Boundary Objects," *Social Studies of Science* 19 (1989): 393.
52 Enqvist et al., "Stewardship as a Boundary Object for Sustainability Research: Linking Care, Knowledge and Agency."
53 Simon West et al., "Stewardship, Care and Relational Values," *Ecosystems and People* 16, no. 1 (January 1, 2020): 4.
54 West et al., "Stewardship, Care and Relational Values," 2.
55 Maria Tengö et al., "Weaving Knowledge Systems in IPBES, CBD and beyond – Lessons Learned for Sustainability," *Current Opinion in Environmental*

Sustainability 26–27 (2017); Albert V. Norström et al., "Principles for Knowledge Co-Production in Sustainability Research," *Nature Sustainability* 3 (January 20, 2020).

56 In another essay in this collection, Sun-Kee Hong acknowledges how native islanders in the Korean archipelago are the best positioned to detect quickly and respond to climate-induced changes to their marine environment. See "The Future of Seascape and the Humanity of Islanders: Focusing on the Korean Archipelago," above 156–174.

57 West et al., "Stewardship, Care and Relational Values," 2.

58 Uno Svedin, "Implicit and Explicit Ethical Norms in the Environmental Policy Arena," *Ecological Economics* 24, no. 2–3 (February 1, 1998).

59 Maria Mancilla Garcia et al., "Towards a Process Epistemology for the Analysis of Social-Ecological Systems" *Environmental Values* 29, no. 2 (April 1, 2020); Simon West et al., "A Relational Turn for Sustainability Science? Relational Thinking, Leverage Points and Transformations," *Ecosystems and People* 16, no. 1 (January 1, 2020).

Bibliography

Amy, Sam R., Matthew S. Heard, Sue E. Hartley, Charles T. George, Richard F. Pywell, and Joanna T. Staley. "Hedgerow Rejuvenation Management Affects Invertebrate Communities through Changes to Habitat Structure." *Basic and Applied Ecology* 16 (2015): 443–51.

Andersson, Erik, Stephan Barthel, Sara Borgström, Johan Colding, Thomas Elmqvist, Carl Folke, and Asa Gren. "Reconnecting Cities to the Biosphere: Stewardship of Green Infrastructure and Urban Ecosystem Services." *Ambio* 43, no. 4 (May 2014): 445–53.

Asah, Stanley T., and Dale J. Blahna. "Motivational Functionalism and Urban Conservation Stewardship: Implications for Volunteer Involvement." *Conservation Letters* 5, no. 6 (2012): 470–77.

Barendse, Jaco, Dirk Roux, Bianca Currie, Natasha Wilson, and Christo Fabricius. "A Broader View of Stewardship to Achieve Conservation and Sustainability Goals in South Africa." *South African Journal of Science* 112, nos. 5–6 (2016): 1–15.

Bennett, Nathan J., Tara S. Whitty, Elena Finkbeiner, Jeremy Pittman, Hannah Bassett, Stefan Gelcich, and Edward H. Allison. "Environmental Stewardship: A Conceptual Review and Analytical Framework." *Environmental Management* 61, no. 4 (2018): 597–614.

Campbell, Lindsay K. "Constructing New York City's Urban Forest: The Politics and Governance of the MillionTreesNYC Campaign." In *Urban Forests, Trees and Greenspace. A Policy Perspective*, edited by Anders L. Sandberg, Adrina Bardekjian, and Sadia Butt, 242–60. New York: Routledge, 2014.

Chapin, F. Stuart, Stephen R. Carpenter, Gary P. Kofinas, Carl Folke, Nick Abel, William C. Clark, Per Olsson, et al. "Ecosystem Stewardship: Sustainability Strategies for a Rapidly Changing Planet." *Trends in Ecology and Evolution* 25, no. 4 (2010): 241–49.

Chapin, F. Stuart, Mary E. Power, Steward T. A. Pickett, Amy Freitag, Julie A. Reynolds, Robert B. Jackson, David M. Lodge, et al. "Earth Stewardship: Science for Action to Sustain the Human–Earth System." *Ecosphere* 2, no. 8 (2011): art89.

Cockburn, Jessica, Georgina Cundill, Sheona Shackleton, and Mathieu Rouget. "The Meaning and Practice of Stewardship in South Africa." *South African Journal of Science* 115, no. 5–6 (2019): 1–10.
Cockburn, Jessica, Eureta Rosenberg, Athina Copteros, S.F. Cornelius, N. Libala, L. Metcalfe, and Benjamin Van Der Waal. "A Relational Approach to Landscape Stewardship." *Land* 9, 224 (2020).
Connolly, James J., Erika S. Svendsen, Dana R. Fisher, and Lindsay K. Campbell. "Organizing Urban Ecosystem Services through Environmental Stewardship Governance in New York City." *Landscape and Urban Planning* 109, no. 1 (2013): 76–84.
Davis, Gary E. "National Park Stewardship and 'Vital Signs' Monitoring: A Case Study from Channel Islands National Park, California." *Aquatic Conservation: Marine and Freshwater Ecosystems* 15 (2005): 71–89.
Enqvist, Johan, Lindsay K. Campbell, Richard C. Stedman, and Erika S. Svendsen. "Place Meanings on the Urban Waterfront: A Typology of Stewardships." *Sustainability Science* 14, no. 3 (2019): 589–605.
Enqvist, Johan P., Simon West, Vanessa A. Masterson, L. Jamila Haider, Uno Svedin, and Maria Tengö. "Stewardship as a Boundary Object for Sustainability Research: Linking Care, Knowledge and Agency." *Landscape and Urban Planning* 179, no. 2 (2018): 17–37.
Enqvist, Johan, Maria Tengö, and Wiebren J. Boonstra. "Against the Current: Rewiring Rigidity Trap Dynamics in Urban Water Governance through Civic Engagement." *Sustainability Science* 11, no. 6 (2016): 919–33.
Folke, Carl, Reinette Biggs, Albert V. Norström, Belinda Reyers, and Johan Rockström. "Social-Ecological Resilience and Biosphere-Based Sustainability Science." *Ecology and Society* 21, no. 3 (2016): art41.
Folke, Carl, Åsa Jansson, Johan Rockström, Per Olsson, Stephen R. Carpenter, F. Stuart Chapin, Anne-Sophie Crépin, et al. "Reconnecting to the Biosphere." *Ambio* 40, no. 7 (2011): 719–38.
Folke, Carl, Henrik Österblom, Jean-Baptiste Jouffray, Eric F. Lambin, W. Neil Adger, Marten Scheffer, Beatrice I. Crona, et al. "Transnational Corporations and the Challenge of Biosphere Stewardship." *Nature Ecology & Evolution* 3, no. 10 (2019): 1396–1403.
Galaz, Victor. *Global Environmental Governance, Technology and Politics: The Anthropocene Gap.* Cheltenham: Edward Elgar, 2014.
Garcia, Maria Mancilla, Tilman Hertz, and Maja Schlüter. "Towards a Process Epistemology for the Analysis of Social-Ecological Systems." *Environmental Values* 29, no. 2 (April 1, 2020): 221–39.
Gupta, Angela S., Samantha Grant, and Andrea Lorek Strauss. "4-H and Forestry Afterschool Clubs: A Collaboration to Foster Stewardship Attitudes and Behaviors in Youth." *Journal of Extension* 50, no. 3 (2012).
Hamilton, Clive. *Growth Fetish.* Sydney: Allen & Unwin, 2003.
Kareiva, Peter, Robert Lalasz, and Michelle Marvier. "Conservation in the Anthropocene: Beyond Solitude and Fragility." *Breakthrough Journal*, February 2012, 29–37.
Kjellén, Bo. *A New Diplomacy for Sustainable Development: The Challenge of Global Change.* London: Routledge, 2014.
Kjellén, Bo. "Diplomacy and Governance for Sustainability in a Partially Globalised World." In *Environmental Values in a Globalising World*, edited by Jouni Paavola and Ian Lowe. London: Routledge, 2004.

Loftus, Timothy T., and Steven E. Kraft. "Enrolling Conservation Buffers in the CRP." *Land Use Policy* 20 (2003): 73–84.

Lokocz, Elizabeth, Robert L. Ryan, and Anna Jarita Sadler. "Landscape and Urban Planning Motivations for Land Protection and Stewardship: Exploring Place Attachment and Rural Landscape Character in Massachusetts." *Landscape and Urban Planning* 99 (2011): 65–76.

Malhi, Yadvinder. "The Concept of the Anthropocene." *Annual Review of Environment and Resources* 42 (2017): 77–104.

Mathevet, Raphaël, François Bousquet, and Christopher M. Raymond. "The Concept of Stewardship in Sustainability Science and Conservation Biology." *Biological Conservation* 217 (2018): 363–370.

McMillen, Heather, Lindsay K. Campbell, Erika S. Svendsen, Kekuhi Kealiikanakaoleohaililani, Kainama S. Francisco, and Christian P. Giardina. "Biocultural Stewardship, Indigenous and Local Ecological Knowledge, and the Urban Crucible." *Ecology & Society* 25, no. 2 (2020): Art.9.

Nicolette, Joseph, Stephanie Burr, and Mark Rockel. "A Practical Approach for Demonstrating Environmental Sustainability and Stewardship through a Net Ecosystem Service Analysis." *Sustainability* 5 (2013): 2152–77.

Norström, Albert V., Christopher Cvitanovic, Marie F. Löf, Simon West, Carina Wyborn, Patricia Balvanera, Angela T. Bednarek, et al. "Principles for Knowledge Co-Production in Sustainability Research." *Nature Sustainability* 3 (January 20, 2020): 182–190.

Österblom, Henrik, Jean Baptiste Jouffray, Carl Folke, and Johan Rockström. "Emergence of a Global Science–Business Initiative for Ocean Stewardship." *Proceedings of the National Academy of Sciences of the United States of America* 114, no. 34 (August 22, 2017): 9038–43.

Rawat, Yashwant S. "Sustainable Biodiversity Stewardship and Inclusive Development in South Africa: A Novel Package for a Sustainable Future." *Current Opinion in Environmental Sustainability* 24 (2017): 89–95.

Schneider, Anselm, Jennifer Hinton, David Collste, Taís Sonetti González, Sofia Valeria Cortes-Calderon, and Ana Paula Dutra Aguiar. "Can Transnational Corporations Leverage Systemic Change Towards a 'Sustainable' Future?" *Nature Ecology and Evolution* 4, no. 4 (2020): 491–492.

Seamer, John H. "Human Stewardship and Animal Welfare." *Applied Animal Science* 59 (1998): 201–205.

Selinske, Matthew J., Benjamin Cooke, Nooshin Torabi, Mathew J. Hardy, Andrew T. Knight, and Sarah A. Bekessy. "Locating Financial Incentives among Diverse Motivations for Long-Term Private Land Conservation." *Ecology and Society* 22, no. 2 (2017): Art.7.

Soulé, Michael. "The 'New Conservation'." In *Keeping the Wild*, 66–80. Washington, DC: Island Press, 2014.

Star, Susan Leigh, and James R. Griesemer. "Institutional Ecology, 'Translations' and Boundary Objects: Amateurs and Professionals in Berkeley's Museum of Vertebrate Zoology, 1907–39." *Social Studies of Science* 19 (1989): 387–420.

Steffen, Will, Paul J. Crutzen, and John R. McNeill. "The Anthropocene: Are Humans Now Overwhelming the Great Forces of Nature?" *Ambio* 36, no. 8 (2007): 614–21.

Steffen, Will, Åsa Persson, Lisa Deutsch, Jan Zalasiewicz, Mark Williams, Katherine Richardson, Carole Crumley, et al. "The Anthropocene: From Global Change to Planetary Stewardship." *Ambio* 40 (2011): 739–61.

Steffen, Will, Katherine Richardson, Johan Rockström, Sarah Cornell, Ingo Fetzer, Elena Bennett, R. Biggs, et al. "Planetary Boundaries: Guiding Human Development on a Changing Planet." *Science* 347, no. 6223 (2015): 1259855.

Subramanian, Meera. "Anthropocene Now: Influential Panel Votes to Recognize Earth's New Epoch." *Nature News*. Accessed September 9, 2021. https://www.nature.com/articles/d41586-019-01641-5.

Svedin, Uno. "Implicit and Explicit Ethical Norms in the Environmental Policy Arena." *Ecological Economics* 24, no. 2–3 (February 1, 1998): 299–309.

Svendsen, Erika S., Lindsay K. Campbell, Dana R. Fisher, James J.T. Connolly, Michelle L. Johnson, Nancy F. Sonti, Dexter H. Locke, et al. "Stewardship Mapping and Assessment Project: A Framework for Understanding Community-Based Environmental Stewardship." Accessed 9 September 2021. https://www.nrs.fs.fed.us/pubs/50447.

Svendsen, Erika S, and Lindsay K Campbell. "Urban Ecological Stewardship: Understanding the Structure, Function and Network of Community-Based Urban Land Management." *Cities and the Environment* 1, no. 1 (2008): 1–32.

Tengö, Maria, Rosemary Hill, Pernilla Malmer, Christopher M. Raymond, Marja Spierenburg, Finn Danielsen, Thomas Elmqvist, and Carl Folke. "Weaving Knowledge Systems in IPBES, CBD and beyond – Lessons Learned for Sustainability." *Current Opinion in Environmental Sustainability* 26–27 (2017): 17–25.

Unnikrishnan, Hita, and Harini Nagendra. "Privatizing the Commons: Impact on Ecosystem Services in Bangalore's Lakes." *Urban Ecosystems* 18 (2015): 613–632.

van Putten, Ingrid, Fabio Boschetti, Elizabeth A. Fulton, Anthony D.M. Smith, and Olivier Thebaud. "Individual Transferable Quota Contribution to Environmental Stewardship: A Theory in Need of Validation." *Ecology & Society* 19, no. 2 (2014): Art.35.

van Riper, Laura. "Women as Collaborative Leaders on Rangelands in the Western United States." *Rangelands* 35, no. 6 (2013): 47–57.

Vega, Dora Carias, and Rodney J. Keenan. "Land Use Policy Agents or Stewards in Community Forestry Enterprises? Lessons from the Mayan Biosphere Reserve, Guatemala." *Land Use Policy* 52 (2016): 255–265.

Welchman, Jennifer. "A Defence of Environmental Stewardship." *Environmental Values* 21 (2012): 297–316.

West, Simon, L. Jamila Haider, Vanessa Masterson, Johan P. Enqvist, Uno Svedin, and Maria Tengö. "Stewardship, Care and Relational Values." *Current Opinion in Environmental Sustainability* 35 (2018): 30–38.

West, Simon, L. Jamila Haider, Sanna Stålhammar, and Stephen Woroniecki. "A Relational Turn for Sustainability Science? Relational Thinking, Leverage Points and Transformations." *Ecosystems and People* 16, no. 1 (January 1, 2020): 304–25.

Index

Abbey, Edward 16
Africa 75–79, 111, 120, 121, 175, 177, 178, 224, 236, 238, 243
African American 76–82, 112, 122
agency 4, 6, 17, 78, 92, 94, 97, 99, 106, 108, 111, 119, 123, 136–138, 142–146, 217, 236, 242, 244–245
agriculture 43–44, 65, 80–81, 83, 105, 108, 137, 144, 147, 179–180, 183–184, 186, 219–220, 222–225, 227–228, 239
Antarctic 159
apocalypse 64, 104–126
aquaculture 162
archipelago 156–169
arid 177–178, 183
Artic 22, 159
Atlantic 76, 79, 111, 123

Bangladesh 226
barley 35–38, 40–44
beaches 18, 34, 37, 89, 109, 111, 112, 235
bear 21
beehives 107
Bible 39, 56
biocultural diversity 165–166
biodiversity 20, 139, 141, 163, 165–166, 212, 218, 227, 234, 238–239, 243
biosphere 17, 63–64, 66, 168, 236, 238, 240–241, 243–244
birds 1, 15–16, 35, 41, 89, 115, 135–136, 138–139, 146–147, 240
boundary 7, 156, 160–161, 162–163, 168, 169, 216–217, 235–237, 241–242, 245
Britain 43, 44, 220

business 2, 7, 14, 17, 31, 40, 42–43, 54, 78–79, 83, 108, 198–200, 202–206, 223, 228, 239

campaign 13–14, 42–43, 205, 238, 241
Canada 41, 87, 128, 198
canals 19, 115
capitalism 6, 22, 29, 33, 42–43, 86, 93, 104, 113–116, 162, 169, 176–177, 185–186, 199, 220, 224–226
carbon dioxide 69, 221
carbon emission 69, 105, 111, 176, 183–186, 218, 221
cars 223, 228
Carson, Rachel 51, 58–62, 63, 65–66, 137
cattle 138, 180–181, 240
chemical 89–90, 105, 196, 198–199
China 108, 218, 219, 224, 226
circular economy 195–197, 199–206
city 13, 18, 19, 31, 108, 110–119, 126–128, 167, 176, 220, 222, 225, 234, 240, 243
climate 6, 7, 20, 32–33, 40, 52, 66, 104–130, 146, 159–161, 164–168, 177–180, 182–184, 186, 205–207, 212, 227, 234
climate change 6, 20, 52, 66–69, 104, 126, 164–166, 168, 182–184, 186, 205–206, 227, 234
colonialism 29, 34, 39, 42–43, 111, 162, 177, 178, 182, 185, 186, 239
communication 4, 7, 108, 160, 161, 163, 205, 218, 222, 231, 242, 245
communist 226
computer 201, 233

Index 253

conscience 38, 60, 121
conservation 3, 13, 17, 22, 51–52, 135–136, 139–141, 142–147, 162, 176, 238–239
conservative 52, 61, 139, 140–141, 143–144
consumption 7, 43, 62, 140–141, 147, 195, 200–206, 219–220, 223–224, 227
cooperation 7, 66, 92, 124, 126, 165, 217–219, 224–225, 231
countryside 52
cow 89, 146
coyote 110, 128, 136–137, 185
crops 33, 35, 43, 75, 81, 82, 176, 177, 185, 219
culture 1, 2, 6, 43, 56, 66, 77–79, 104, 119, 121, 156–157, 159–169, 212–213, 227–228, 230, 243

degradation 6, 127, 181–183, 186, 241
desert 29, 146, 177–179, 183, 185
development 42–43, 50–51, 57, 64, 76, 89, 91, 106–107, 109, 117, 122, 135, 138, 140–141, 147, 156–158, 161–162, 164–170
drought 177, 179–182

Earth 1–3, 5–7, 16, 21, 32, 34–38, 44, 50, 52–58, 60–66, 88, 92, 94, 98, 100, 105, 108, 118, 124–126, 159, 175, 220
earthquake 36–38, 89, 168
ecology 15, 17–23, 50, 52–63, 65–66, 76, 116, 159, 161–162, 164–166, 168, 175, 182, 201, 230, 236–240, 243–245
economy 5, 14, 17, 19, 29–33, 42, 60, 63, 66–67, 80–83, 86–87, 105, 107–111, 114, 135, 141–143, 145, 147, 156–164, 168–169, 176–177, 186, 212–213, 219, 222–230, 237–238, 243
ecosystem 17, 21, 56–63, 66, 105, 109–112, 118, 124–125, 137, 139, 159, 161–166, 168, 175–176, 178, 183, 204–206, 213, 236, 238–239
endangered 3–4, 6, 88, 110, 121, 135, 137, 139–140, 142, 146–147
energy 17, 22, 63, 64, 137–142, 146, 166, 199, 200, 203, 219–221, 223–224, 227–229
ethnicity 98
eugenics 86, 91

Europe 31, 43, 56, 111, 162, 175, 179, 197, 202–203, 220
evolution 1, 7, 118–126, 157, 162, 176–177, 212–213
extermination 136
extinction 109, 120, 122, 125, 135, 137, 147

factory 43, 176, 203, 220
farmers 136, 138, 141, 144–145, 223, 243
fauna 1, 2, 115, 135, 137
fish 1, 19, 37, 114, 139, 141–147, 159, 161–162, 164–169
fishery 162, 165, 167
flood 37, 64, 104, 107, 110, 115, 117
flora 1, 2, 115, 135, 137
flowers 56, 57, 65, 115
forest 6, 18, 59, 63–43, 110, 112, 136–137, 142, 176–178, 181–182, 215

garden 20, 32–33, 36, 39–40, 76, 97, 107, 240
geography 2, 15–16, 75, 176
global 17, 20, 63, 67, 69, 108, 111, 117, 118, 120–121, 123, 125, 159–160, 166, 168, 178, 179, 182, 184, 195, 201, 212–213, 215, 218–219, 223–229, 234–235, 241, 243–245
global warming 15, 212
goat 35, 175
government 2, 3, 13–14, 21, 32, 43, 106, 108, 135, 139, 141, 143–144, 166, 183, 198, 206, 213, 218, 226–227, 236, 238–239, 243
grassland 175–176, 179, 182–184
grazing 136, 177, 181, 189
greenhouse gases 214, 227

habitat 64, 66, 135, 140, 142, 147
habitation 40, 41, 111
harvest 40–41, 87, 109, 114, 161, 169
herbal remedies 107
herders 175, 178, 240
herds 181, 183
history 1–2, 4–6, 31, 33, 59, 62, 77, 92, 95, 97, 104, 106–107, 109–111, 113–114, 117–118, 158–159, 165–166, 177, 179, 183, 185, 186, 219, 228, 230, 239, 245
homes 15, 31, 32, 37, 52, 79, 81, 87, 91, 94, 97, 117, 164, 223, 228

254 Index

horse 43, 137–138, 175, 181
human race 86, 88–89, 91–92, 94–95, 99
humanist 161
hunting 41, 136–137, 144–145
hydroponics 107
hypothesis 67, 168, 215, 217

identity 15–16, 18–20, 23, 76, 95, 157–160, 163–164, 167, 240, 242
industry 3, 13, 63, 67, 76, 105, 108, 137–142, 146, 157, 165, 167, 179, 183, 196–201, 212, 220–224, 227–229, 234–235, 241
insecticides 60–61, 63, 90
insects 109, 135
international 89, 105, 113, 146, 159, 168, 178, 215, 235
Iran 226
Islam 226

knowledge 2, 4, 6, 14, 18, 22–23, 29, 32–33, 37, 39, 41, 53, 58–59, 63, 65, 87, 89, 105, 156–158, 160–169, 177, 182–183, 204, 212–216, 219, 224–225, 227–230, 236, 242–245
Korea 6, 156, 162, 164–165, 168–170, 224

landfill 165, 199–201
landscape 2, 6, 13–16, 18–23, 34, 53, 59, 76, 78, 80, 110, 112, 114–118, 159, 161, 165–166, 176–177, 179, 181, 185, 186, 236, 238, 240
Leopold, Aldo 5, 21, 51, 58–61, 136
liberal 51, 86, 212, 226, 228, 239
litter 201, 237
livestock 136, 138, 175–176, 178, 181–184, 189
local 6, 13–15, 20–22, 38, 39, 54, 63, 76, 78, 79, 108, 110, 159, 166, 178, 182, 184, 186, 213, 214, 218–219, 221–223, 225–226, 230, 234, 240, 242

mammal 135
management 1, 4, 7, 13, 19, 22, 59–61, 80, 86, 88, 91, 96, 99–100, 106, 136–137, 142–146, 178, 180, 181–184, 186, 195–198, 200–203, 206, 222–223, 225, 236–240
marine 111, 158–159, 161–165, 167–168, 201, 206, 236, 241
Marx, Karl 29–30, 42–43, 176, 213

meadows 19, 180
moral 22, 30, 32, 52, 98, 117, 118, 122, 124, 146, 213, 227, 237, 239
mountains 14–16, 18, 20, 58, 59, 143, 145
mud 119, 159, 167

nuclear power 63–64, 87, 89, 90, 105, 127, 212, 218, 227

ocean 2–4, 64, 107, 115–117, 158–159, 162, 164–165, 167, 201, 236, 238, 240–241, 243–244

Pacific 43, 126, 162, 168
park 3–5, 13–15, 18–22, 77, 137, 146, 234–235, 237
pasture 65, 180–181, 185
placental economy 6, 93–95, 97, 99
planes 223, 228
plains 18, 177, 179, 181, 183
plantation 6, 75–84
plastic 64, 162, 199, 201, 206, 242
pollution 33, 62, 64, 88–89, 105, 128, 195–199, 201, 205–207
population 59, 62–63, 65, 90, 97, 137–138, 145, 147, 161, 167, 169, 176–177, 201, 212, 220–230, 237–238
post-human 94, 126
power 6, 20, 23, 32, 37, 38–39, 50–51, 54, 58, 63–64, 78, 81–83, 86–87, 105–109, 111–113, 115, 119–126, 135, 137–138, 142–143, 147, 156, 158, 162, 175–178, 185, 203, 214–215, 222–223, 226, 240–241, 244
prairie 15, 18, 175, 179, 182
predators 21, 81–83, 110, 116, 136–137, 139, 183, 186
productivity 220–222

race 34, 55, 87–92, 95, 99, 165
radiation 89
rainfall 176, 178–179, 181, 182
rainforest 64, 176, 215
ranchers 136–138, 140–141, 145–146, 182
re-greened 179
religion 2, 4, 5, 32, 38, 51–53, 56, 58, 60, 62, 66, 104–105, 107
reproduction 86–100, 225, 228

resources 2, 3, 17, 18, 22, 35–36, 38, 42, 51, 57, 64, 65, 94, 106, 108–109, 112–113, 118, 135–138, 142, 144, 156–157, 159, 161–168, 175–176, 178, 195, 199, 200, 206, 212–213, 218–219, 221–225, 227–229, 235–236, 239
rice 42
risk 63, 91, 124, 196, 198, 200, 202–203, 205, 212, 218–219, 222, 227, 229, 239
rural 15, 21, 52–58, 76, 138, 141, 143, 224

Saudi Arabia 226
savannahs 120, 175, 189
science 3–4, 17, 37, 56, 58, 60, 62–63, 93, 104–105, 111, 140–141, 147, 161–162, 167, 177–183, 185, 220, 235, 241–243
scrublands 175, 182, 189
seashores 15
sharecroppers 75, 81–83
sheep 34, 175
slave 2, 6, 39, 75–76, 78–83, 87–88, 93, 95, 97, 99, 108–111, 113–115, 117–123
society 17, 30–31, 33, 52, 57, 62, 106, 135, 156–159, 161, 164–166, 168, 176, 198–199, 202, 218–219, 223, 225–236
soil 1, 2, 4, 35, 43, 55, 75, 80, 176, 179–180, 183–184, 186
species 1–4, 6–7, 15, 17, 20, 24, 56, 61, 63, 65, 86, 88–89, 92, 95, 98, 102, 107–110, 120, 122, 124, 128, 135, 137–147, 212, 214–215, 228, 237–238

spirituality 22, 36, 51, 58, 62, 65–66, 104–105
steps 127, 185
subsistence 108, 113, 125, 178, 221–225, 227–229
sustainability 6, 7, 17, 19, 23–24, 51, 57, 108, 125, 156, 160, 164–169, 177, 195, 198–204, 227, 230, 234–239, 241–242, 244–245
swamp 18, 104, 110, 112

Talmud 54
technology 62, 76, 82, 87, 91, 92, 95, 99, 102, 105, 119, 122–123, 160–162, 167, 169, 200, 212–213, 215–219, 221, 229, 242
tide 159, 165, 167, 169
topography 65, 156–161, 163–165, 167
tourism 13, 14, 164–165, 169
toxic 89, 90, 234
tundra 127, 175

urban 2, 15, 21, 60, 76, 107, 143, 145, 163, 212–213, 219–223, 224, 226–229, 236, 238, 240

vegetation 59–60, 80, 176, 179–182, 184

waste 3, 7, 57, 64, 90, 93, 162, 176, 195–197, 199–206, 219, 221, 223–224, 227, 234
wheat 44
wilderness 4, 19, 39, 110, 137
wildlife 16, 22, 24, 63, 135–147, 183
wolf 15, 21, 58, 59, 110, 114, 128, 136–139, 141–143, 145
women 6, 32, 88–90, 92–95, 97–99

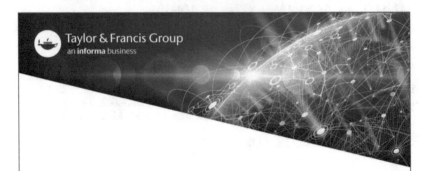